Learn Data Science Using SAS Studio

A Quick-Start Guide

Engy Fouda

Apress®

Learn Data Science Using SAS Studio: A Quick-Start Guide

Engy Fouda
Hopewell Junction, NY, USA

ISBN-13 (pbk): 978-1-4842-6236-8
https://doi.org/10.1007/978-1-4842-6237-5

ISBN-13 (electronic): 978-1-4842-6237-5

Managing Director, Apress Media LLC: Welmoed Spahr
Acquisitions Editor: Susan McDermott
Development Editor: Laura Berendson
Coordinating Editor: Rita Fernando

Cover designed by eStudioCalamar

Cover image designed by Freepik (www.freepik.com)

Distributed to the book trade worldwide by Springer Science+Business Media New York, 1 New York Plaza, New York, NY 10004. Phone 1-800-SPRINGER, fax (201) 348-4505, email orders-ny@springer-sbm.com, or visit www.springeronline.com. Apress Media, LLC is a California LLC and the sole member (owner) is Springer Science+Business Media Finance Inc (SSBM Finance Inc). SSBM Finance Inc is a **Delaware** corporation.

For information on translations, please e-mail booktranslations@springernature.com; for reprint, paperback, or audio rights, please e-mail bookpermissions@springernature.com.

Apress titles may be purchased in bulk for academic, corporate, or promotional use. eBook versions and licenses are also available for most titles. For more information, reference our Print and eBook Bulk Sales web page at http://www.apress.com/bulk-sales.

Any source code or other supplementary material referenced by the author in this book is available to readers on GitHub via the book's product page, located at www.apress.com/9781484262368. For more detailed information, please visit http://www.apress.com/source-code.

Printed on acid-free paper

To my daughter, Areej, and my husband, Hesham, a big thank you to both of you from my deep heart. I would have never finished this book without your support and encouragement. You created for me a happy and safe life while the world outside was crazy and losing its sanity. I am a lucky person to have you both in my life. Thank you so much! Love you both!

To my mom, Suzan, and dad, Samir, I miss both of you madly and wish you were with me celebrating this book. I am sure that you are celebrating and happy together in the heavens. I owe you everything and hope you are as proud of me as you always were. Till we meet, I love you.

Finally, to my brothers, Haitham and Khaled, you never gave up on me and always believed in me. I always need your trust that I can achieve anything. I appreciate it and with it I gain my ability to move forward. I hope you like this book.

Table of Contents

About the Author

Engy Fouda is an author, freelance engineer, and journalist. Currently, she teaches SAS, Docker Fundamentals, Docker for Enterprise Developers, Docker for Enterprise Operations, and Kubernetes at several venues as a freelance instructor. She is an Apress and Packt Publishing author. She works as a freelance journalist and publishes her work at various media outlets. She holds two master's degrees, one in journalism from Harvard University, Extension School, and another in computer engineering from Cairo University, Egypt. Moreover, she earned the Data Science Professional Graduate Certificate from Harvard University, Extension School. She has taught academically as a teacher assistant at the German University in Cairo and the American University in Cairo. She volunteers as the team lead for Momken Group (Engineering for the Blind), Egypt Scholars Inc. The team designs and manufactures devices and develops Arabic applications for the visually impaired people in the Middle East and North Africa region. Also, she volunteers as a member-at-large and the newsletter editor of the IEEE Mid-Hudson Section. She has published several books that made Amazon's best-seller charts for Arabic books.

About the Technical Reviewer

 **Allan Bowe** is a SAS geek with a passion for HTML5 apps on SAS. Allan has made a number of contributions to the SAS community, such as SASjs (an adapter for bidirectional communication between HTML5 and SAS), sasjs-cli (a command-line tool for managing SAS project compilation, build, and deployment), and macrocore (a SAS macro library for building SAS apps on both SAS 9 and Viya).

When not building web apps, Allan is working on Data Controller, a commercial data capture, data quality, and data governance web app for both SAS 9 and Viya.

Introduction

The book's scope is primarily to introduce SAS Studio, a free data science web browser–based product for non-commercial and academic usage. SAS Studio is also known as SAS University Edition. The power of SAS Studio relies on its visual point-and-click user interface that generates SAS code. Users can create data analysis reports without writing a line of code, unlike with R and Python. Hence, data cleaning, statistics, and visualization are easy to do.

The book's case study analyzes the presidential elections data in Maine, which is part of a project I did at Harvard University. Chapter 1 explains the case study in more detail.

In addition to the presidential elections, the book uses real-life examples like analyzing stocks, oil and gold prices, crime, marketing, and healthcare. The whole book follows the paradigm of data science in action to demonstrate how easy it is to perform complicated tasks and visualizations in SAS Studio.

The book starts from scratch in step-by-step, hands-on labs, and includes screenshots of every step.

It will provide readers the required expertise in data science and analytics using SAS Studio, such as how to do the following:

- Import and export raw data files

- Manipulate and transform data

- Combine SAS data sets

- Create summary statistics and reports using SAS procedures

- Identify and correct data, syntax, and programming logic errors

- Compare between samples using T-test and Anova

- Predict new values using linear regression

- Create visualizations

Moreover, it will show how to do visualizations, including maps, step-by-step. In many cases, readers will not need to write a line of code, because SAS has a powerful graphical user interface. It is much easier to learn compared to R and Python. However, every example will explain the auto-generated code and how to edit it to perform more-complicated advanced tasks. The book will introduce you to multiple SAS products, such as SAS Studio, SAS Viya, SAS Analytics, and SAS Visual Statistics.

I teach most of these contents as a course at one of the Microsoft Partner Centers. The students always get amazed by how much they learn in merely a few days.

Who Should Read This Book?

The primary audience of this book is students who are newbies to data science and might not have deep programming experience. Consequently, university professors can use it as a handout for their courses. Also, technical instructors, like me, who teach professionals from various industry sectors at certified training centers can teach from it.

Also, system analysts and scientists who are experienced but new to SAS will find faster and more efficient tools to achieve their daily tasks. From my experience, I learned from the attendees of my courses that many government agencies migrated to SAS. Moreover, data journalists and investigative reporters will find the book easy to follow and will be able to generate pretty visualizations quickly.

How Is This Book Organized?

In general, the book tries to balance between using SAS point-and-click and the code. The users might be tempted to rely on the integrated development environment (IDE). However, the book uses the IDE merely to introduce the users to the various tasks, along with explaining the code so as to be able to do advanced tasks.

The book has three parts. Part I, "Basics" (Chapters 1, 2, and 3), gets you familiar with the SAS interface and the basic essential tasks. Chapter 1 focuses on drawing a general idea of the case study of the presidential election project in the state of Maine and its outputs. Throughout the book, you will learn how to get those output charts and analytics step-by-step.

Then, Part II, "More Programming" (Chapters 4, 5, and 6), focuses on more advanced programming aspects. Part III, "Advanced Topics" (Chapters 7, 8, and 9), takes you from analyzing historical data to predicting the future and introducing you to more advanced SAS platforms.

Finally, in Chapter 9, I try to give some insights from my personal experience on how to get certified and how to make money online through teaching, writing, and competing online. Moreover, I shall give you more details on the data science graduate professional certificate at Harvard University, Extension School.

All the SAS example code is in the "Example Code" folder, and datasets required for this book are in the "Datasets" folder of this book's source code. Go to http://www.apress.com/9781484262368 and click the Download Source Code button to access it.

PART I

Basics

CHAPTER 1

Data Science in Action

In this chapter, we will introduce the case study of the book, which analyzes voters' data in the state of Maine. It is based on a project I did at Harvard University in 2016 during my master's degree. In fall 2016, the project for my "A Practical Approach to Data Science" course was to predict the presidential election results in every state. The project was under the guidance and supervision of Professor Larry Adams, who set the project milestones and requirements. I was responsible for forecasting Maine's outcome for the 2016 and 2020 elections.

The project was done in two phases. The first was to predict the results for the 2016 election. After verifying our data and results against what actually happened in the election, the second phase started. It was to include the new data that was generated in 2016 and use it to predict the results of the year 2020. Therefore, some charts and exercises in this book include 2016 data. Whenever possible, I collected any related historic data. For the prediction, I used historic election data going back to 1960.

I defined voters' groups by age, gender, education, demographics, and race. After studying the state from reliable academic sources, I identified issue categories like the economy, education, the environment, health care, and gun control.

Similarly, I listed the state's issues that would influence the presidential election by using the county ballot topics. Using the voting patterns of each party since 1960, poll accuracy, and the electoral votes, I tried different prediction methods and algorithms, such as Monte Carlo and Bayes, and statistical testing, such as T-test, chi-square, and others. Afterward, I had to compare my results to other forecast sites, like Five-Thirty-Eight. My prediction was correct for 2016.

This project was an exciting experience in which I converted cognitive features to numbers and crunched them to come up with results. Similarly, through other data science projects, I learned how to predict outcomes so as to drive decision making based upon measuring trends and studying patterns.

© Engy Fouda 2020
E. Fouda, *Learn Data Science Using SAS Studio*, https://doi.org/10.1007/978-1-4842-6237-5_1

Data Science Process

The data science process starts with forming a question or hypothesis, then collecting relevant raw data, then cleaning and exploring that data, then modeling and evaluating, then deploying, visualizing, and communicating results in reports, as shown in Figure 1-1.

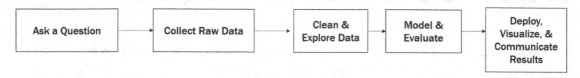

Figure 1-1. *Data science process*

Questions vary according to the field; for example:

- Politics: Will Trump win in Maine in 2016 and 2020?

- Facebook: How can you make people stay on Facebook longer?

- Medical: Is this tumor cancer or not?

- Hospital Management: How can you decrease patients' wait lines so as to increase patients' satisfaction?

The second step is collecting raw data. For example, in the politics question: Will a particular candidate win in a certain state?

Collecting all the voters' information—age, race, education, income, gender, and industry—is a crucial step, as is collecting the ballot data and voting results from over the years. The more historical data we have, the more accurate our predictions are. Furthermore, we should collect information on the population distribution throughout the years.

The third step is cleaning this raw data, from managing the missing values, outliers, repeated rows, and misspelled information, to adjusting the columns' data types, unifying the format of the values, and so on.

The fourth step is trying several models and comparing their results with each other, depending upon the problem's nature. In the presidential election problem, I used Monte Carlo and Bayes algorithms.

The fifth and final step is visualizing the results and communicating them in plain language in our reports. This step is the primary goal of the whole process because it holds the predictions to the answer to the first question that initiated the whole process.

Case Study: Presidential Elections in Maine

As I mentioned in the previous section, the data science process starts with a question. In this project, my question is: Will Donald Trump win in the state of Maine in the 2016 and 2020 presidential elections?

Population

The second step is collecting as much related data as possible. Therefore, I started with the population.

From information on the population distribution over Maine's counties, found at the U.S. Census Bureau, I learned that it is not uniformly distributed. There are vast areas that are either unpopulated or that have only one person living in them. While the red dots in the south look small, more than 5,000 people live in each of them. Therefore, I should not be deceived by the maps distributed by the presidential campaigns or by the mainstream media.

The following logical step was to get the voters' information. Some states publish their voters databases for free, and anyone could download them. However, in Maine, this was not the case. The state sold the voter databases to the political parties. So, I contacted the Secretary of State.

The office replied that to obtain voters files and updates from Maine's Central Voter Registration system, the requesting person or entity must be from the following five cases:

1. A candidate or person or entity working on a candidate's campaign

2. Someone working for a party

3. A person or entity involved in a referendum campaign that will be on the ballot in Maine in the next statewide election

4. A person or entity involved in specific get-out-the-vote efforts in Maine (the efforts have to be identified, including name, location, and date of events in Maine)

5. An individual who has been elected or appointed to and currently serving in a municipal, county, state, or federal office, but only for use for the official's authorized activities, not to turn over to another entity

The cost was based on the number of records obtained; the fee was scheduled in Title 21-A, section 196-A. A statewide voter file, which contained almost one million records, was $2,200.

After a few emails back and forth explaining that I needed them for a research project and sending some verifications, the office kindly sent me for free a DVD with all the required information, hiding the unneeded data like last names and so on.

The first table on the DVD has the voters' information and is shown in Figure 1-2. The columns are first name, year of birth, enrollment code, special designations, date of registration, congressional district, county ID, changed date, and date of last statewide election with VPH.

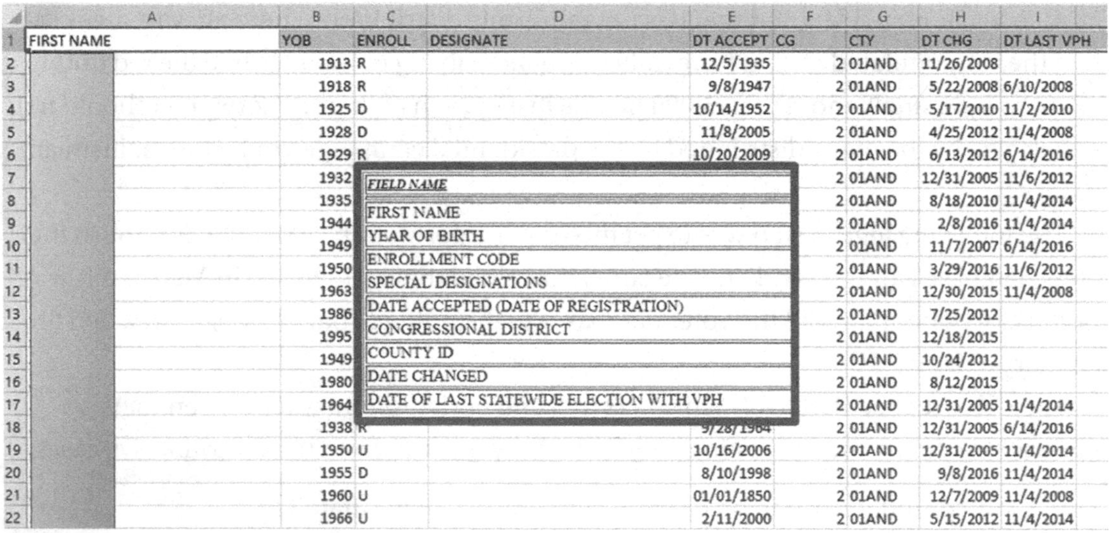

	A	B	C	D	E	F	G	H	I
1	FIRST NAME	YOB	ENROLL	DESIGNATE	DT ACCEPT	CG	CTY	DT CHG	DT LAST VPH
2		1913	R		12/5/1935	2	01AND	11/26/2008	
3		1918	R		9/8/1947	2	01AND	5/22/2008	6/10/2008
4		1925	D		10/14/1952	2	01AND	5/17/2010	11/2/2010
5		1928	D		11/8/2005	2	01AND	4/25/2012	11/4/2008
6		1929	R		10/20/2009	2	01AND	6/13/2012	6/14/2016
7		1932		*FIELD NAME*		2	01AND	12/31/2005	11/6/2012
8		1935		FIRST NAME		2	01AND	8/18/2010	11/4/2014
9		1944		YEAR OF BIRTH		2	01AND	2/8/2016	11/4/2014
10		1949		ENROLLMENT CODE		2	01AND	11/7/2007	6/14/2016
11		1950		SPECIAL DESIGNATIONS		2	01AND	3/29/2016	11/6/2012
12		1963		DATE ACCEPTED (DATE OF REGISTRATION)		2	01AND	12/30/2015	11/4/2008
13		1986		CONGRESSIONAL DISTRICT		2	01AND	7/25/2012	
14		1995		COUNTY ID		2	01AND	12/18/2015	
15		1949		DATE CHANGED		2	01AND	10/24/2012	
16		1980		DATE OF LAST STATEWIDE ELECTION WITH VPH		2	01AND	8/12/2015	
17		1964				2	01AND	12/31/2005	11/4/2014
18		1938	R		9/28/1964	2	01AND	12/31/2005	6/14/2016
19		1950	U		10/16/2006	2	01AND	12/31/2005	11/4/2014
20		1955	D		8/10/1998	2	01AND	9/8/2016	11/4/2014
21		1960	U		01/01/1850	2	01AND	12/7/2009	11/4/2008
22		1966	U		2/11/2000	2	01AND	5/15/2012	11/4/2014

Figure 1-2. *Voters' information*

The second table contains a registered and enrolled voters report, as in Figure 1-3. The columns of this table are the county name, municipality name, ward precinct, congressional district, state senate, county commissioner district, the party, and the total. The parties listed in the file are Democratic, Green Independent, Libertarian, Republican, and unenrolled.

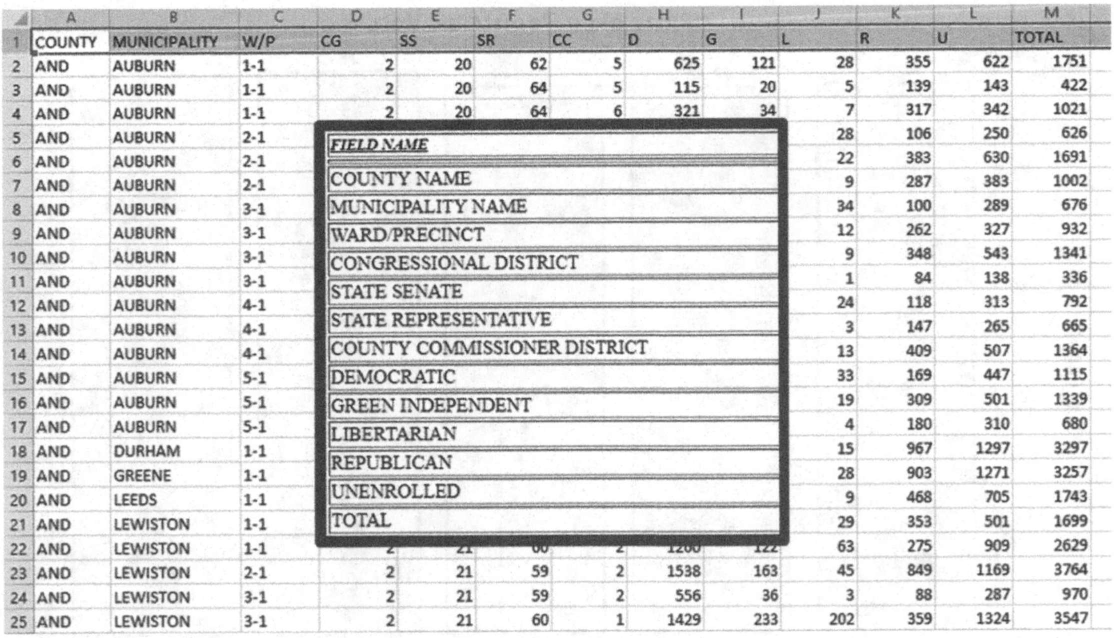

COUNTY	MUNICIPALITY	W/P	CG	SS	SR	CC	D	G	L	R	U	TOTAL
AND	AUBURN	1-1	2	20	62	5	625	121	28	355	622	1751
AND	AUBURN	1-1	2	20	64	5	115	20	5	139	143	422
AND	AUBURN	1-1	2	20	64	6	321	34	7	317	342	1021
AND	AUBURN	2-1							28	106	250	626
AND	AUBURN	2-1							22	383	630	1691
AND	AUBURN	2-1							9	287	383	1002
AND	AUBURN	3-1							34	100	289	676
AND	AUBURN	3-1							12	262	327	932
AND	AUBURN	3-1							9	348	543	1341
AND	AUBURN	3-1							1	84	138	336
AND	AUBURN	4-1							24	118	313	792
AND	AUBURN	4-1							3	147	265	665
AND	AUBURN	4-1							13	409	507	1364
AND	AUBURN	5-1							33	169	447	1115
AND	AUBURN	5-1							19	309	501	1339
AND	AUBURN	5-1							4	180	310	680
AND	DURHAM	1-1							15	967	1297	3297
AND	GREENE	1-1							28	903	1271	3257
AND	LEEDS	1-1							9	468	705	1743
AND	LEWISTON	1-1							29	353	501	1699
AND	LEWISTON	1-1	2	21	00	2	1200	122	63	275	909	2629
AND	LEWISTON	2-1	2	21	59	2	1538	163	45	849	1169	3764
AND	LEWISTON	3-1	2	21	59	2	556	36	3	88	287	970
AND	LEWISTON	3-1	2	21	60	1	1429	233	202	359	1324	3547

Overlay legend (FIELD NAME box):

FIELD NAME
COUNTY NAME
MUNICIPALITY NAME
WARD/PRECINCT
CONGRESSIONAL DISTRICT
STATE SENATE
STATE REPRESENTATIVE
COUNTY COMMISSIONER DISTRICT
DEMOCRATIC
GREEN INDEPENDENT
LIBERTARIAN
REPUBLICAN
UNENROLLED
TOTAL

Figure 1-3. *Registered and enrolled voters report*

This raw data was messy and contained many wrong values and outliers. For example, the age of one voter was 220 years, while his date of birth states that he was about 67 years old at that time. Some voters' information was missing, and so on. Again, as mentioned earlier, always clean your data: outliers, missing data, adjust data formatting, and explore your data.

Not only that, but also you should collect as much historical data as you can. So, I started digging and collected as much data as I could find. From the United States Census Bureau, I downloaded more tables (http://www.census.gov/topics/public-sector/voting/data/tables.html).

I grouped the voters by gender, age, and race.

Gender

As shown in Figure 1-4, I found that the registered female population is larger than the registered male population in Maine. However, the percentage difference for both genders is less than ±2%. As for how the candidates should use this information, the campaigns' representatives could wear the cancer awareness ribbon to play on women's compassion, as the women's turnout was always higher than the males'. This recommendation shows how data and numbers can control not only the speech topics but even what the representatives wear.

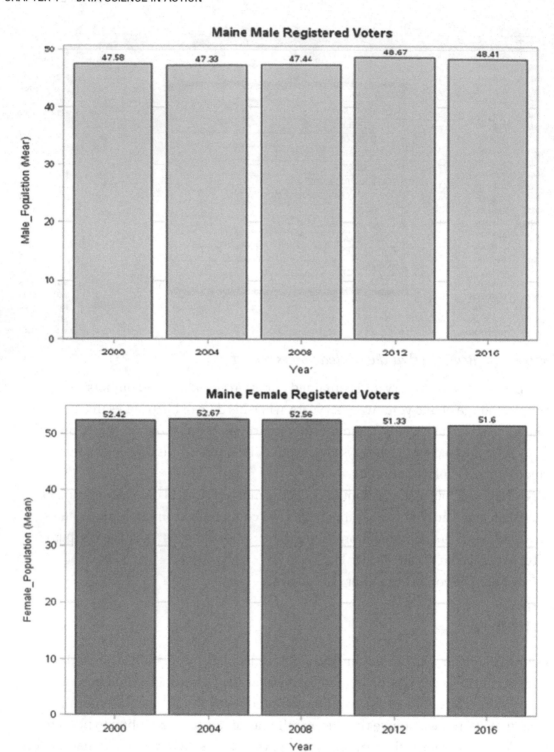

Figure 1-4. *Maine registered voters grouped by gender*

Race

Regarding race in Maine, more than 96 percent of the population is white (as shown in Figure 1-5). Therefore, I grouped the black, Asian, and Hispanic voters as non-white registered voters.

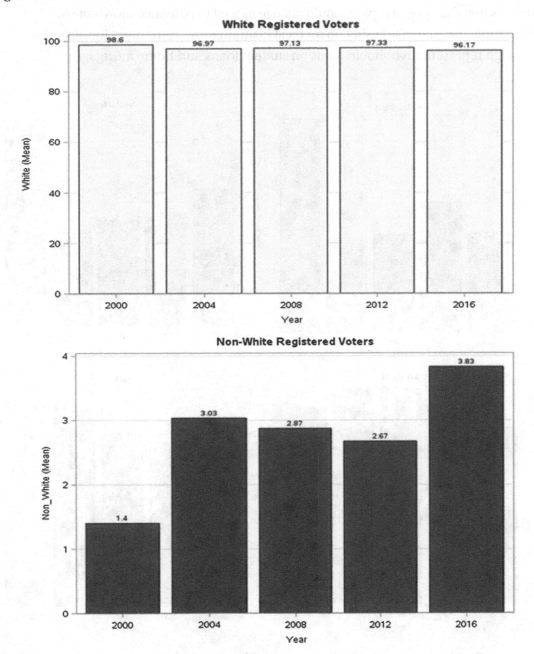

Figure 1-5. *Maine registered voters grouped by race*

Age

In 2016, there was a remarkable decrease in the number of registered voters who were over the age of sixty-five, as seen in Figure 1-6 (note that the scales are different). On the other hand, there was an increase in the 18–24 and 25–44 age groups. This finding indicates that the speech topics should change as well to convince more voters. For example, instead of focusing on medical insurance and retirement funding, the campaign representative should focus on student loans and home mortgages.

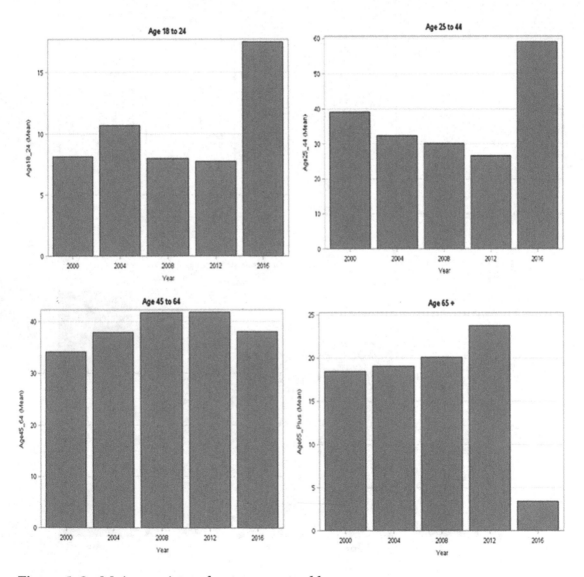

Figure 1-6. *Maine registered voters grouped by age*

Voter Turnout

I checked the voter turnout for past years to see how many voters got dressed and went out in Maine's snowy streets to vote. Figure 1-7 shows voter turnout from the years 2000 to 2016 as percentages, and that the winning party was the Democratic Party for those years. If you try this exercise with a swing party, the columns' colors will change to reflect the winning party.

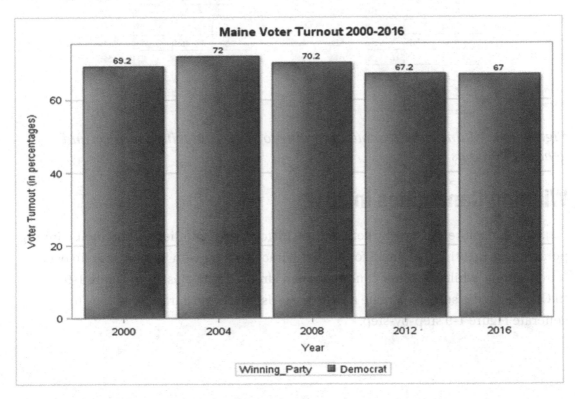

Figure 1-7. *Maine voter turnout 2000–2016*

Again, I inspected the voter turnout in recent years, according to the winning party and with age, race, and gender groupings. In Figure 1-8, the D represents the Democratic Party, the R stands for the Republican Party, and the O is for all other parties combined.

		2004			2008			2012			2016	
	D	R	O	D	R	O	D	R	O	D	R	O
Voter Turn Out	396,842	330,201	8,069	421,923	295,273	10,636	401,306	292,276	9,352	357,735	335,593	38,105
%	53.57%	44.58%	1.09%	57.71%	40.38%	1.45%	56.27%	40.98%	1.31%	47.80%	44.90%	5.10%
18 to 24	48%	50%	1%	71%	26%	3%	65%	30%		50%	42%	5%
25 to 29	48%	50%	1%	62%	36%	2%	60%	35%		46%	44%	8%
30 to 39	48%	50%	1%	60%	38%	2%	59%	36%		40%	50%	7%
40 to 49	59%	39%	1%	54%	44%	2%	51%	44%		48%	46%	6%
50 to 64	59%	39%	1%	61%	36%	3%	58%	40%		47%	48%	4%
65+	54%	45%	1%	45%	53%	2%	55%	43%		56%	39%	3%
Male	48%	49%	1%	52%	46%	2%	50%	46%		41%	52%	6%
Female	57%	42%	1%	64%	34%	2%	64%	34%		55%	39%	5%
White	53%	45%	11%	58%	40%	2%	57%	40%		47%	46%	5%
Non-White										56%	33%	10%

Figure 1-8. *Maine voter turnout according to party and with age, race, and gender groupings*

Winning Candidates in 2012

I started to explore the data by generating a histogram of candidates and their winning percentages in 2012 in Maine. I found that Barack Obama won by a big margin over Mitt Romney, while the other candidates had almost zero votes, as in Figure 1-9. In Chapter 3, the section "Create a Histogram Using a Bar Chart" shows how to generate Figure 1-9 step-by-step.

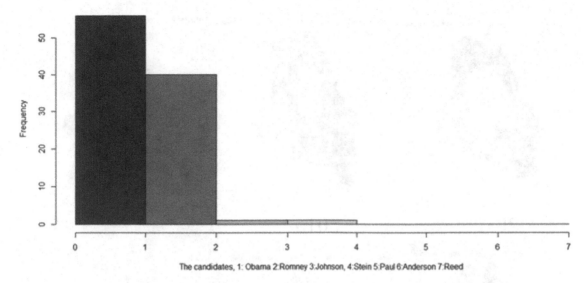

Figure 1-9. *Histogram of candidates and their winning percentages in 2012*

Next, I tried to visualize the data on the map, per county. Figure 1-10 shows the Democratic counties with the darkest shade and the Republican Party as a lighter shade for the years 2000–2016. However, remember that in Maine the area does not correspond to the population. Chapter 8 lists the steps and the codes needed to generate the maps in Figure 1-10.

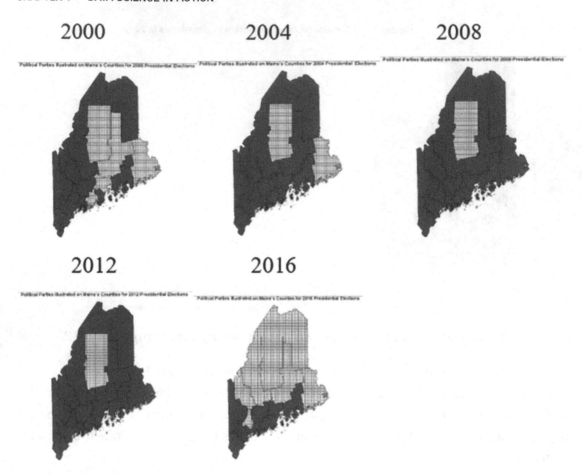

Figure 1-10. *Map of how Maine's counties voted in the presidential elections in years 2000–2016*

Categories/Issues

All presidential election debates focus on the issues and how every candidate is going to deal with them. Based upon these strategies, voters might change their minds and vote based on a specific issue. Therefore, it is crucial to invistigate the issues.

Issues Facing the Country

In 2016, according to CNN, Maine voters thought that the most crucial issues facing the country were foreign policy, immigration, economy, and terrorism. However, these issues are not the issues that face Maine itself.[1] From the map, Maine is not affected by foreign policy.

---

[1] Data source: http://www.cnn.com/election/results/exit-polls/maine/president

From the race statistics, it is clear that Maine has almost no immigrants. I searched briefly for any terroristic attacks in Maine and found nothing. The only lasting common issue all over the world was the economy. Hence, I started searching for Maine's issues in the ballots.

Categories: Issues Facing Maine

On the Ballotpedia website, the ballots are listed as in Table 1-1.

Table 1-1. *The Issues Facing Maine Residents*[2]

2000	2004	2008	2012	2016
Assisted Death	Education	Bonds	Marriage	Marijuana
Business	Taxes	Taxes	Education	Education
Lottery	Hung & Fish	Gambling	Water	Firearms
Taxes		Water	Transportation	Economics
Suffrage		Transportation	Bond Issues	Elections
LGBT				Bonds
Fishing				

I found that there are specific issues that are not resolved over the years; for example, fishing, bonds, water, education, and taxes. I deduced that Maine has water and tax problems. Again, I recommended that the speeches of the presidential campaigns should include a proposed solution or a plan for these specific issues. The speech that tackles these issues might win over Maine's voters.

From the information I collected and by surveying the local newspapers, I predict the major five issues in Maine in 2020 to be as follows:

1. Education

2. Economics

3. Bonds

4. Health care

5. Taxes

---

[2]Data sources: https://ballotpedia.org/Maine_2000_ballot_measureshttps://ballotpedia.org/Maine_2004_ballot_measureshttps://ballotpedia.org/Maine_2008_ballot_measureshttps://ballotpedia.org/Maine_2012_ballot_measureshttps://ballotpedia.org/Maine_2016_ballot_measures

Factors Affecting Maine's Economy

As we talk about the economy, looking for income sources is crucial. According to the *Portland Press Herald* and Maine's revenue datasets provided on Maine.gov website,[3] the income sources are as follows:

1. Agriculture

2. Forestry

3. Fishing

4. Hunting

5. Taxes

6. Management of companies and enterprises

7. The finance and insurance sector

Comparing the Unemployment Rate of Maine to That of the Rest of the States

Donald Trump promised to offer more jobs and to return manufacturing firms to the United States instead of outsourcing them. Many news outlets claimed that his jobs promise is what made the Rust Belt states inclined to vote for him instead of Hillary Clinton. So, I searched for the unemployment rate in Maine and compared it to the rest of the country to check if Trump's promise would work with Maine voters or not. I found that Maine did not have an unemployment issue in 2016. Hence, I predicted that Trump's promise would not be as appealing to Maine's voters as it would be in some other states.

Median Income for Maine

To be more specific and for research purposes, I looked up the median income for Maine; then, I compared it to the median income of the United States. According to Table 1-2, from 2005 until 2015, Maine's median income was slightly less than that of the United States. However, there was an increase in the three years before the elections.

---

[3]Data source: `https://www.pressherald.com/2019/03/26/maine-incomes-up-4-3-percent-in-2018/`

Table 1-2. *Historical Real Median Household Income for Maine[4]*

Date	US	Maine
2018	$61,937	$55,602
2017	$61,807	$57,649
2016	$60,291	$55,542
2015	$59,116	$54,578
2014	$56,969	$52,515
2013	$56,415	$50,718
2012	$56,288	$51,180
2011	$56,507	$51,507
2010	$57,762	$52,878
2009	$58,921	$53,657
2008	$60,829	$54,459
2007	$61,601	$55,710
2006	$60,490	$54,233
2005	$59,604	$55,168

How Did Exit Polls Show the Income Factor in the 2016 Election in Maine?

Looking from another perspective, according to CNN, the exit polls showed that Maine would mostly vote for Clinton except voters whose salaries ranged from $30,000 to $50,000.[5] These voters would vote for Trump. However, they represented only 19 percent of the sample. Hence, the majority in Maine would vote for Clinton.

---

[4]Data source: http://www.deptofnumbers.com/income/maine/
[5]Data source: http://www.cnn.com/election/results/exit-polls/maine/president

Reviewing Past Elections to See if There Are Any Predictable Outcomes (Patterns)

Finally, the last step of the project was to collect the historical data for the past elections and use this data for the next step of modeling and prediction. As I mentioned earlier, I neglected any other party and focused solely on Democrats and Republicans because the rest got almost zero percentages in the 2012 elections.

Table 1-3. *The Historical Results of Past Presidential Elections in Maine[6]*

Year	D	D%	R	R%
2016	**357,735**	**47.80%**	335,593	44.90%
2012	**401,306**	**56.27%**	292,276	40.98%
2008	**421,923**	**57.71%**	295,273	40.38%
2004	**396,842**	**53.57%**	330,201	44.58%
2000	**319,951**	**49.10%**	286,616	44.00%
1996	**312,788**	**51.60%**	186,378	30.80%
1992	**263,420**	**38.80%**	206,820	30.40%
1988	243,569	43.90%	**307,131**	**55.30%**
1984	214,515	38.80%	**336,500**	**60.80%**
1980	220,974	42.30%	**238,522**	**45.60%**
1976	232,279	48.07%	**236,320**	**48.91%**
1972	160,584	38.50%	**256,458**	**61.50%**
1968	**217,312**	**55.30%**	169,254	43.10%
1964	**62,264**	**68.84%**	118,701	31.16%
1960	**181,159**	**42.95%**	240,608	57.05%

---

[6]Data source: https://bangordailynews.com/2016/11/09/politics/elections/clinton-leads-maine-but-trump-poised-to-take-one-electoral-vote/

Seeing the data collected in Table 1-3, I was surprised to find out that historically Maine was a Republican state, and then it became a strong Democratic state. Hence, I decided to describe Maine as a "Lean Democrat" state because Democrats won seven out of fifteen times from 1960 to 2016. I am predicting that it will mostly vote Democrat in 2020 with a small margin. Since 1992, Democrats have won straight in Maine. However, if the Democratic Party does not pay more attention to Maine, it might switch to be a Republican one. According to *Bangor Daily News*, Clinton did not visit Maine after September 2015 and sent surrogates instead. For the first time in years, the Democratic Party lost one electoral vote to the Republican Party. (Bangor is a city in Maine.)

Modeling

For modeling, it is always better to try different algorithms for prediction and compare their results, and not rely on only one model. For this project, I used Monte Carlo and Bayes algorithms.

For the statistical tests, I used the following methods:

- Histograms

- Box plots

- Proportion test

- T-test

- Decision tree

- Chi-square test

- Scatter plot and linear regression

My 2016 Predictions

After running the Monte Carlo algorithm in SAS University Edition using the past election results, the output was as shown in Figure 1-11.

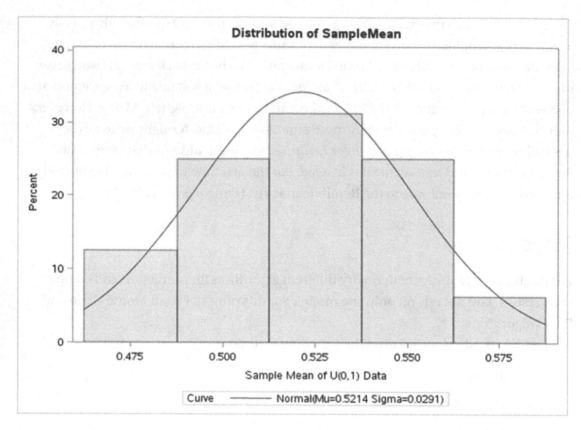

Figure 1-11. *The 2016 prediction results of Monte Carlo algorithm*[7]

Figure 1-11 shows that Hillary Clinton would win Maine by a percentage mean of 52.14 percent. Actually, she won by 48 percent only. So, the simulation detects that she won Maine, and she did. Therefore, I got the full mark in this project and an A in the course. Hurray!

My 2020 Predictions

I predict that the Democratic Party will win three of Maine's four electoral votes by 50.8 percent, as shown in Figure 1-12.

---

[7]Data source: `https://ballotpedia.org/Presidential_election_in_Maine,_2016`

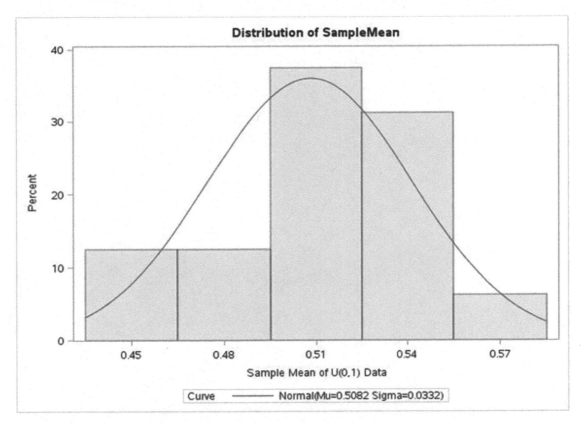

Figure 1-12. *The 2020 prediction results of Monte Carlo algorithm*

The likelihood probability for Maine is that the Democrats will win 2020 by a minor margin of about +3% over the Republicans, where the historical data from 1960 to 2016 shows that the mean percentage of Democrats is about 43 percent while the mean percentage of the Republicans in Maine is 40 percent. It is the same difference as there was between the Democrats and Republicans in 2016 in Maine.

If the Democratic Party is not careful, however, Maine might switch to the Republican Party.

Summary

This chapter focuses on drawing the main ideas of the case study of the presidential elections project in the state of Maine, as well as its charts. In the next chapter, we shall see how to install SAS Studio and write our first SAS program.

CHAPTER 2

Getting Started

SAS Studio enables you to import and export raw data files, manipulate and transform data, combine SAS data sets, create summary reports using SAS procedures, identify and correct data, and debug syntax and programming logic errors. Moreover, it enables you to do visualizations, including maps, in easier steps than other programming languages such as R or Python. However, it has a programming editor as well to customize and write your own code if needed.

How Do You Install SAS Studio?

The installation comprises four steps. You can follow them step-by-step at the following URL: `https://www.sas.com/en_us/software/university-edition/download-software.html`.

As the installation steps can change, it is better to consult this URL for the latest updated online guide.

What Is SAS and SAS Studio?

SAS stands for Statistical Analysis System, and SAS Studio is its interface. SAS Studio has a powerful integrated development environment (IDE), which includes both the tasks' graphical user interface (GUI) and the code editor. If you use the GUI, which uses the point-and-click approach, SAS Studio auto-generates the equivalent code for you. Or, you can use the code editor to write your own code.

Throughout the book, we will go through both approaches: using the GUI and writing code. For the examples that use the GUI, we will explain the generated code. For some advanced examples or user-customized ones that can be done solely by code, I shall explain the required code line by line, along with the procedures and their options.

© Engy Fouda 2020

E. Fouda, *Learn Data Science Using SAS Studio*, https://doi.org/10.1007/978-1-4842-6237-5_2

It is worth mentioning that SAS Studio also has two modes: the SAS Programmer and the Visual Programmer. The Visual Programmer mode uses the drag-and-drop method. You can change the mode from the drop-down menu in the upper bar on the right. However, the book focuses on the SAS Programmer mode.

Tour

The SAS Studio interface is basically divided into two vertical panes. The first one is the navigation pane, which has several tabs; for example, Server Files and Folders, Tasks and Utilities, Snippets, and Libraries. The second one is the work area, which displays the programs' editor/code, log, and results.

The Server Files and Folders tab contains all the folders and subfolders that are inside the shared folder that you have created on your machine. Remember that SAS Studio runs inside a virtual machine; hence, it does not see your operating system files and folders by default. It sees only what is inside that shared folder. Later in this chapter, you will learn several ways to import files from your operating system to the virtual machine where SAS Studio runs and vice vera. After doing so, you will be able to use the files inside SAS Studio.

The Tasks and Utilities tab has task GUIs that you can use within the SAS Studio IDE. Later in this chapter and the book, you will learn how this tab eases your data science process, and how, with a few mouse clicks, it helps you to finish most of the essential steps to explore, clean, and process your data and help you to make your predictions.

The Snippets tab has some of the most-used tasks as ready-made code for you. You can reuse this code for your datasets and customize them. You will learn more about it later in this chapter.

The Libraries tab is where the datasets, that SAS Studio uses to execute the tasks and utilities, are located.

Figure 2-1. *SAS Studio interface*

After you successfully install the SAS Studio inside the virtual machine, click on Libraries in the left pane. Then click on SASHELP, then CARS. You will have a table viewer, as in Figure 2-1, displaying a dataset that has information about cars. The middle pane shows the columns of that dataset. By default, "Select all," highlighted with red in Figure 2-1 to the left, is selected. Now, uncheck it and check "Make" and "Model" only. The right pane will show the data of these two columns only. Try to resize them. You can drag any of the columns to the left or to the right to rearrange them. You can sort any column ascending or descending by clicking on the column name.

Now, click on the icon highlighted in Figure 2-1 in the middle. This icon is the Expand/Collapse button. The first time you click on it, the left pane will collapse, giving you an expanded view of the task that you are working on. If you click on it a second time, the pane will be back.

Another way to view the data is by using code. Let us write our first SAS code. The third icon from the left, when you click on it, will display a drop-down menu, as shown in Figure 2-2.

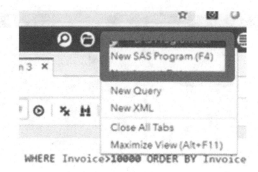

Figure 2-2. *Open a new SAS program*

Click on New SAS Program and start typing "pr," as in Figure 2-3. You will find that helping menus pop up, showing the keywords to autocomplete what you are typing, along with an explanation and additional links for examples and documentation, as shown in Figure 2-3. This feature is useful for beginners.

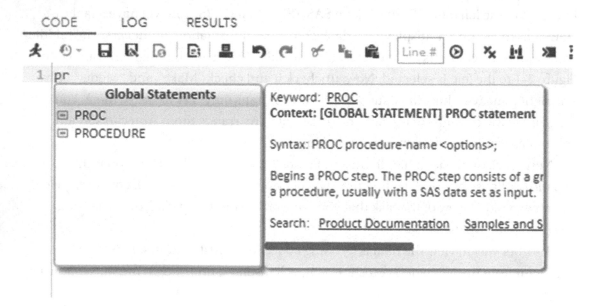

Figure 2-3. *Autocomplete feature and helpful pop-up menus*

However, when you get more experienced and familiar with SAS Studio and would like to disable it, you can do that by clicking on Preferences, as shown in Figure 2-4.

Figure 2-4. *Preferences*

Now, click on Code and Log and uncheck "Enable autocomplete (CTRL + spacebar or Command + spacebar)," as shown in Figure 2-5. As this book targets the beginners, we will not change the defaults.

Figure 2-5. *Code and Log to enable or disable code autocomplete*

Now, type in the code shown in Listing 2-1.

Listing 2-1. Proc Print

```
proc print data=SASHELP.CARS;
run;
```

Click the Run button, as shown in Figure 2-6. Don't forget the semicolon and run; to end the procedure section.

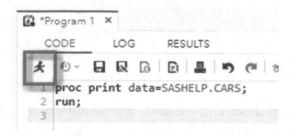

Figure 2-6. *Run button*

The results will be prepared to be printed. You can print your results as HTML, PDF, or RTF by clicking on the highlighted icons, shown in Figure 2-7.

Obs	Make	Model	Type	Origin
1	Acura	MDX	SUV	Asia
2	Acura	RSX Type S 2dr	Sedan	Asia
3	Acura	TSX 4dr	Sedan	Asia
4	Acura	TL 4dr	Sedan	Asia
5	Acura	3.5 RL 4dr	Sedan	Asia
6	Acura	3.5 RL w/Navigation 4dr	Sedan	Asia
7	Acura	NSX coupe 2dr manual S	Sports	Asia

Figure 2-7. *Print your results as HTML, PDF, or RTF*

You just wrote, successfully executed, and printed the report for your first SAS program. Hurray! Congratulations!

Now, click on Libraries in the left pane. Then, click on SASHELP, then CARS. To get the code that SAS Studio auto-generated to display the table, click on the icon highlighted in Figure 2-8.

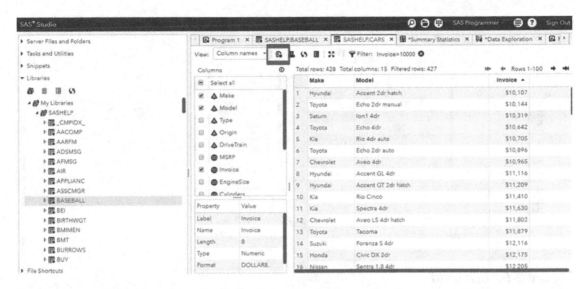

Figure 2-8. *The icon that displays the code that creates the current table*

In the coming chapters, we shall use this feature to modify the code that SAS Studio generates for us to perform more-advanced tasks. For example, in Chapter 3, in the section concerning the bubble chart, there is an example where you will edit the generated code to do some customization to the chart's appearance.

Tasks

Let us make an easy task. Listing data is a third way to view your data within the SAS Studio interface. Click on Tasks ➤ Data ➤ List Data. In DATA, choose SASHELP.CARS. In List Variables, choose the following variables: Make, Model, Invoice. In Group Analysis By, choose Origin, as in Figure 2-9.

Figure 2-9. *List data*

Then, click Run. Figure 2-10 shows the output of the List Data example.

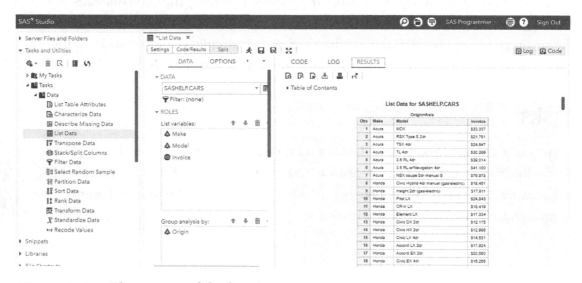

Figure 2-10. *The output of the list data*

Reports

To continue the tour and see how easily you can generate reports in SAS Studio, click on the Tasks and Utilities tab ➤ Tasks ➤ Summary Statistics, as in Figure 2-11.

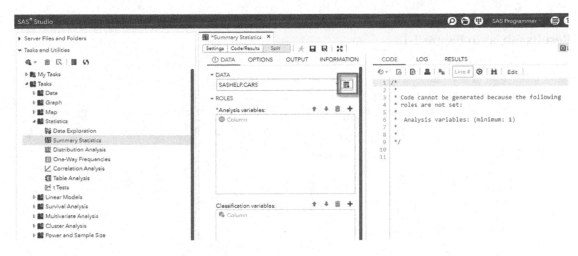

Figure 2-11. *Summary Statistics*

In Data, click on the table. Select SASHELP.CARS. In the Analysis Variables, select Weight, as in Figure 2-12. Instantly, after you select Weight, observe that SAS Studio auto-generated the code for you in the most right pane, as in Figure 2-12. In the first line of code, SAS Studio uses PROC MEANS to compute the summary statistics, the dataset is SASHELP.CARS, then some options to the PROC MEANS as the standard deviation, mean, minimum, maximum, and the number of observations. Later in the book, we will learn how to change these options and customize them. The second line is the analysis variable, which is Weight. Finally, it ends the procedure with RUN and semicolon.

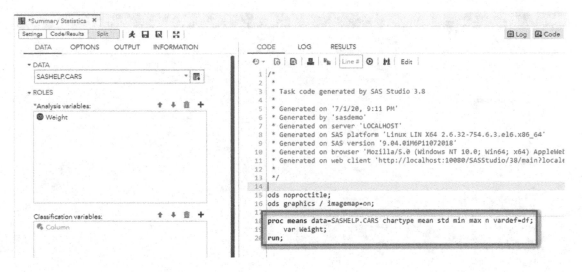

Figure 2-12. *Observe the code*

Now, click Run to get the generated report. The reports are always displayed in the Results tab, as in Figure 2-13.

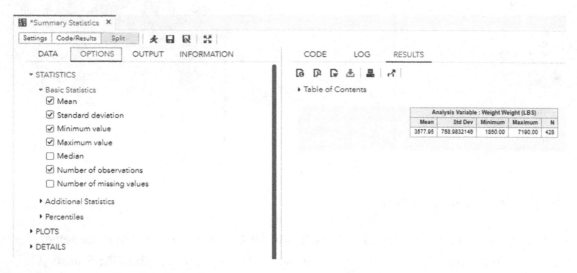

Figure 2-13. *The Results tab*

Graphs

Let's make a quick graph. Click on Tasks and Utilities ➤ Tasks ➤ Graph ➤ Bar Chart, as in Figure 2-14.

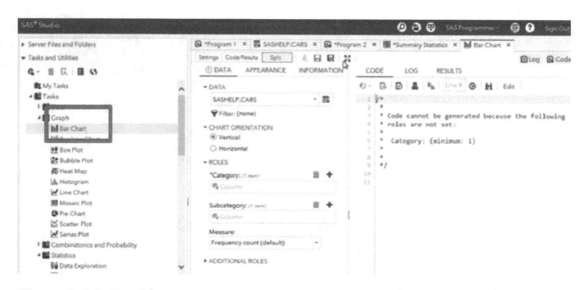

Figure 2-14. *Bar Chart*

Now, as in Figure 2-15, change Data to SASHELP.FISH.

1. Select Species.

2. Measure should be Variable.

3. Select Weight.

4. Then change Statistic to Mean.

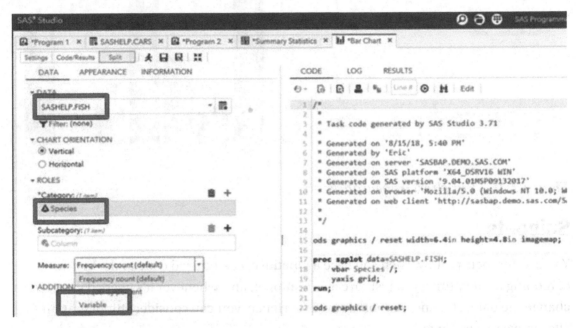

Figure 2-15. *Do the following changes*

As in Figure 2-16, change the title to "Average Weight for Different Species." In Bars, check "Set Color" and change the color. In Details, change Effect to Sheen. Then, click Run. Figure 2-16 shows the output.

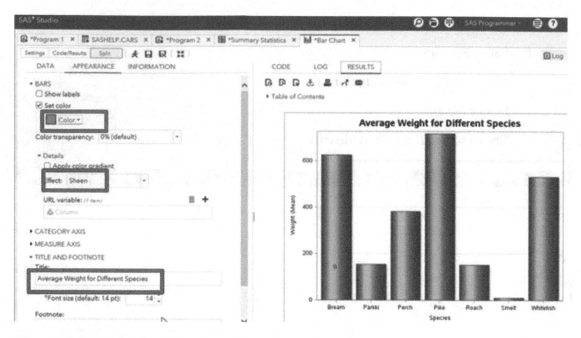

Figure 2-16. *Bar chart Appearance tab*

Snippets

You use snippets mainly when you have a repetitive, customized code to do. Instead of redoing it every time, you just save it as a snippet, and whenever it's needed you just change the dataset name, variables, and title. Hence, you can consider snippets as an alternative to custom tasks.

So, What Are Snippets?

- It's ready-made SAS code.

- You can customize it.

- You can add your own snippet.

Let's use two examples to demonstrate how to use the ready-made snippet and another to create your own new snippet.

Example to Use Ready-Made Snippet

- Click Snippets ➤ Graph.

- Select Hbar. The code is as shown in Figure 2-17.

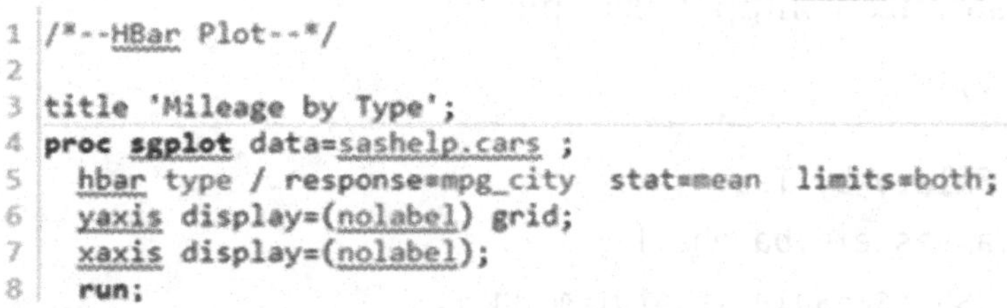

```
1  /*--HBar Plot--*/
2
3  title 'Mileage by Type';
4  proc sgplot data=sashelp.cars ;
5    hbar type / response=mpg_city  stat=mean  limits=both;
6    yaxis display=(nolabel) grid;
7    xaxis display=(nolabel);
8    run;
```

Figure 2-17. *HBar plot snippet*

The code in Listing 2-2 is in the "Example Code" folder, from which you can copy and paste it.

Listing 2-2. Snippet Example

```
/*--HBar Plot--*/

title 'Mileage by Type';
proc sgplot data=sashelp.cars ;
  hbar type / response=mpg_city  stat=mean  limits=both;
  yaxis display=(nolabel) grid;
  xaxis display=(nolabel);
  run;
```

- Edit the code to be as shown in Figure 2-16, as follows:

  - Edit the title.

  - Change the dataset to baseball instead of cars.

  - Change hbar from "type" to "team."

  - Change response from "mpg_city" to "salary."

  - Remove limits.

The modified code for Listing 2-3 is located in the "Example Code" folder, from which you can copy and paste it. Its screenshot looks like Figure 2-18.

Listing 2-3. Customized Snippet Example

```
/*--HBar Plot--*/

title 'Average Salary by Team';
proc sgplot data=sashelp.baseball ;
  hbar team / response=salary  stat=mean  ;
  yaxis display=(nolabel) grid;
  xaxis display=(nolabel);
  run;
```

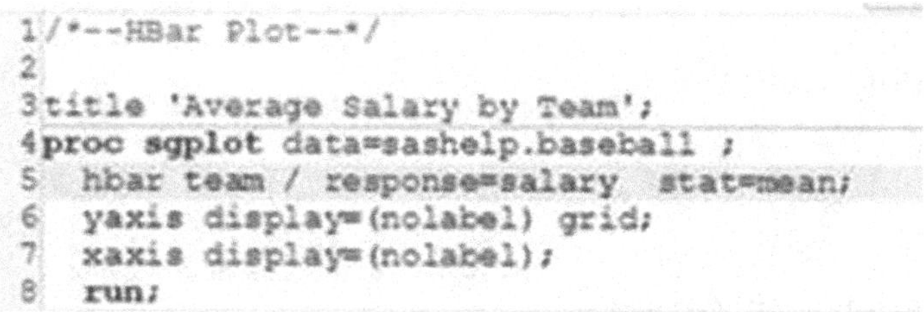

```
1/*--HBar Plot--*/
2
3title 'Average Salary by Team';
4proc sgplot data=sashelp.baseball ;
5  hbar team / response=salary  stat=mean;
6  yaxis display=(nolabel) grid;
7  xaxis display=(nolabel);
8  run;
```

Figure 2-18. *Customize the snippet to your new dataset and variables*

- Run the code.

To Create Your Own Snippets

As in Figure 2-19, click on the Code tab, and then choose the icon for (Add to My Snippets).

```
ram 1  ×   SASHELP.FISH  ×    Program 2  ×    *Program 3  ×   *P

Settings   Code/Results   Split           ⚡ 🖫 🖫 ⛶

   CODE          OG       RESULTS

 ⏱ ▾   🗔   🗔   🖩   🖩   Line #   ⓞ   ⟷   Edit

  1          Add to My Snippets
  2
  3
  4
  5
  6  ods graphics / reset width=6.4in height=4.8in imagemap;
  7
  8
  9  proc sgplot data=  SASHELP.BASEBALL ;
 10
 11
 12
 13      title height=14pt "Average Salaries by Team";
 14
 15  hbar Team
 16  /
 17  response=Salary           stat=mean      ;
 18
 19
 20      xaxis
 21              grid                    ;
 22
```

Figure 2-19. *Add to My Snippets*

Main Components of a SAS Program

A SAS program consists of two steps: the data step and the proc step. Each program can contain as many data and proc steps as you need, as in Figure 2-20.

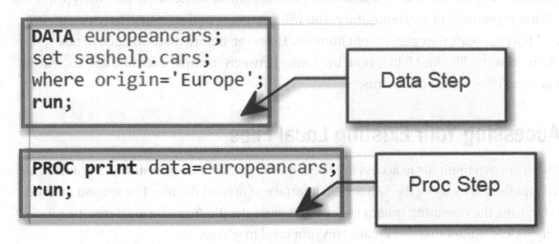

Figure 2-20. *Main components of a SAS program*

Data Step

The data step is where you prepare your data for processing by defining the dataset, combining datasets, cleaning your data, creating new variables, checking for errors, removing any variables, and so on.

Variable Types

In SAS, there are two variable types: numeric and character. SAS dates and times are not standard numeric or character variables. They have their functions but are stored as numbers. In data science, the terminology that data scientists use is different. The continuous variables or the quantitative variables are represented in numeric variables, while the categorical or qualitative variables are represented in character variables.

Proc Step

The proc step is where you define your procedures to process clean data and generate reports. You use the proc step to summarize, sort, list, filter, or perform any other task from the Tasks tab.

Libraries

The next stop in our tour is the third tab in the left pane, Libraries. Libraries are where you load your tables, so SAS Studio can access the tables' data. There are two types of libraries: permanent and temporary. You will learn how to distinguish between them and how to create new permanent libraries. However, before that, we need to talk about how the files and folders can be shared between the host machine and the virtual machine where SAS Studio runs.

Accessing Your Existing Local Files

There are two methods to access the files on your machine from inside SAS Studio. The first method is by using the SAS Studio interface to upload the file. The second method is by using the operating system to copy and paste the file from the local machine to the "myfolders" folder that you created on your local machine.

The First Method

Click on Server Files and Folders ➤ Upload. Then, choose the Excel or CSV or any format file from your local machine to upload it into SAS Studio (see Figures 2-21 and 2-22).

Figure 2-21. *Click the Upload icon*

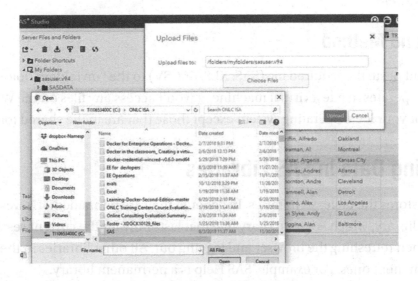

Figure 2-22. *Select the file from server files and folders*

Similarly, if you want to download a file from inside SAS Studio to your local machine, select the file and click the Download icon to download the file (see Figure 2-23). You will find it in your "Downloads" folder on your local machine (see Figure 2-23).

Figure 2-23. *Download icon to download any file from SAS Studio to your operating system*

The Second Method

Just copy and paste the required file (SAS, XLS, or CSV) to the "myfolders" folder. As SAS Studio operates inside a virtual machine, it won't access any files on the Windows filesystem or your local operating system except those that are in the shared folder.

Accessing Data in SAS Libraries

SAS Studio stores the tables in libraries. In general, they are permanent. However, there is the Work library, which is a temporary one that lasts only for the current session and vanishes when refreshing the browser and logging out. All other libraries in the Libraries tab are permanent ones. For example, SAS Help is a permanent library.

How Do You Distinguish Between Permanent and Temporary Libraries?

Any dataset has a two-level name separated by a dot, with the following syntax: <first level name>.<second level name>; for example, sashelp.cars or newlib.adults. This is an indication that your dataset is saved in a permanent library. However, if your dataset has "work" as its first level, it is saved in a temporary library; for example, work.ratedata.

SAS Studio uses the Work library by default. Hence, for any dataset that has only one level, "work" will be its first level, and it is temporary; for example, income. You will find it saved under the Work library in the Libraries tab. So, work.income is similar to income; both are the same and indicate a temporary library called income.

Create a New Library

There are two ways to create new libraries: by code or by GUI. Creating a new library by code is only one command: libname <Library Name> <Path>;

Library Naming Conventions

The library that you create must have a name between one and eight characters in length, beginning with a letter or underscore. It cannot start with a number or contain any special characters. The remaining characters of the name can be letters, numbers, and underscores. No special characters are allowed.

The path can be any folder saved in the Server Files and Folders tab. Type the code shown in Listing 2-4, found in the "Example Code" folder, from which you can copy it and paste it into a new program to create your new library, and then click run.

Listing 2-4. Create a New Library Using libname Statement

```
libname newlib '/folders/myfolders';
```

You will find your library listed in the Libraries tab. If not, click Refresh.

The second method is by using the GUI. Click on the Libraries tab, and then click the New Library icon, as in Figure 2-24.

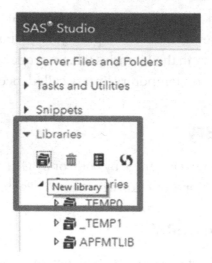

Figure 2-24. *Create a new library using the GUI*

Select your folder, as in Figure 2-25.

Figure 2-25. *Select the location of your library*

Enter its name and path, as in Figure 2-26. It is better to check "Re-create this library at start-up," so you can find it whenever you refresh SAS Studio or restart the virtual machine. Otherwise, you will not find it and will need to re-add it to your libraries list manually.

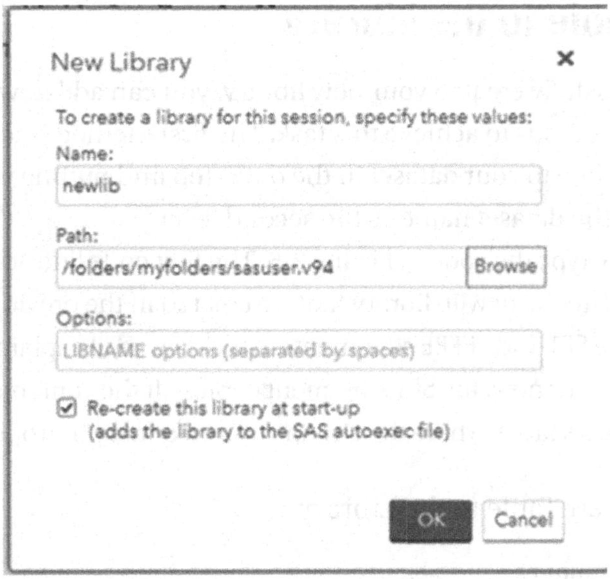

New Library   ×

To create a library for this session, specify these values:

Name:

newlib

Path:

/folders/myfolders/sasuser.v94   Browse

Options:

LIBNAME options (separated by spaces)

☑ Re-create this library at start-up
   (adds the library to the SAS autoexec file)

OK   Cancel

Figure 2-26. *Creating the new library options*

Click OK. You will find your new library in the Libraries list, as in Figure 2-27.

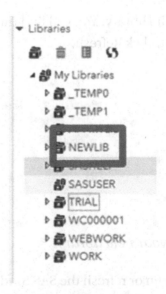

Figure 2-27. *The new library in the Libraries list*

Add a New Table to the Library

After you have successfully created your new library, you can add new datasets or tables to it. There are two methods to achieve this task. The first method is done by code by adding a two-level name to your dataset in the data step and putting your library name as the first level and the dataset name as the second level.

In a new program, type the code in Listing 2-5. The first code line adds a new table called "europeancars" to the newlib library that we created in the previous step. The second and third lines use the SET and WHERE statements, which we shall explain in detail later in the book in Chapter 6. For now, the SET statement copies all the contents of sashelp.cars to the newlib.europeancars table, where the Origin column equals "Europe" only.

Listing 2-5. Add New Table to the Library

```
data newlib.europeancars;
set sashelp.cars;
where origin='Europe';
run;
```

Click Run. Click on the newlib library. You will find the dataset listed under it, as in Figure 2-28. If you did not find it, click Refresh.

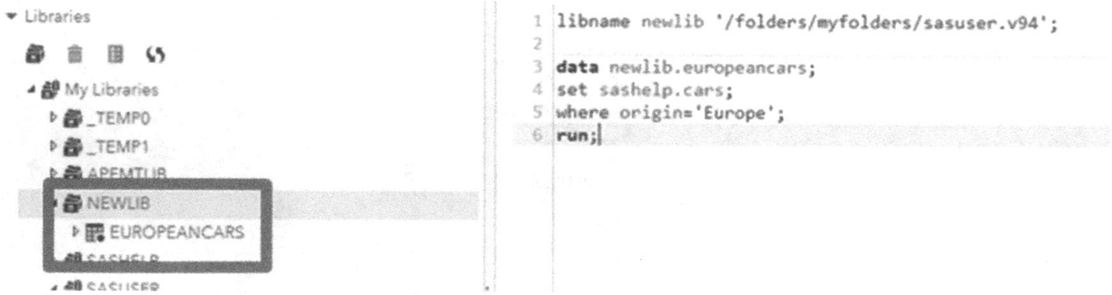

Figure 2-28. *Save a table to your new library*

Now, login to SAS Studio again or refresh the SAS Studio window. You should find the library and the table after signing in.

The second method is by importing the dataset. Select Files and Folders ➤ Import Data ➤ select the file name, as in Figure 2-29. Then click Run.

Figure 2-29. *Import data*

Click on Select Files, as in Figure 2-30.

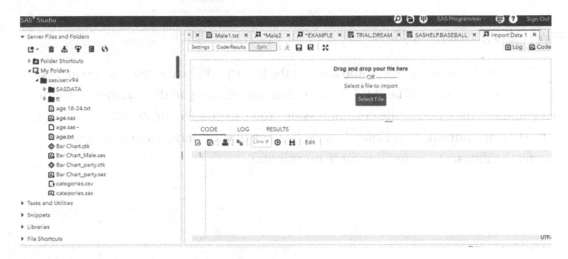

Figure 2-30. *Select our file from the Server Files and Folders tab*

Then, choose the file that you would like to import, as in Figure 2-31.

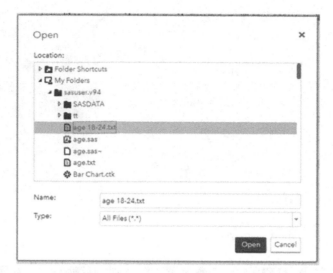

Figure 2-31. *Select the file*

The default values in SAS Studio are that the library will be the temporary one, Work, and the dataset name will be IMPORT<Number> depending on the number of imports that you have done during the current session. You can change the defaults by clicking on the Change button, as in Figure 2-32. Select one of your permanent libraries and rename the dataset to something more meaningful.

Figure 2-32. *Change the table and library names*

Moreover, if you scroll down, you will find more options to change, like the delimiter and the first row, as in Figure 2-33.

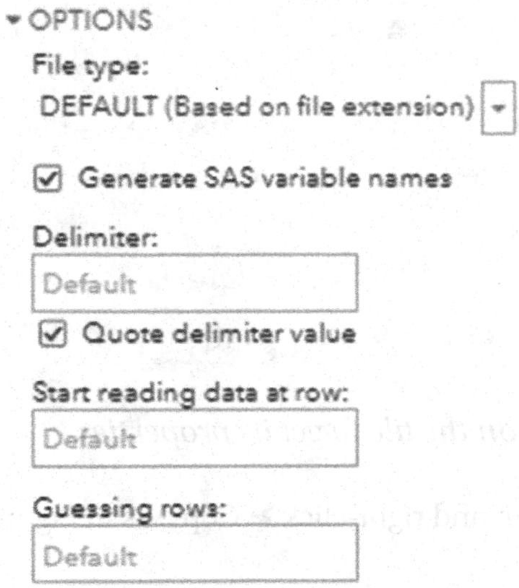

Figure 2-33. *Import data options*

The auto-generated SAS variables' names will be extracted from the first row in your data file. If your data file's first row is data and not header names, you can uncheck that. In the code, in the data step, you can rename the variables using the Rename keyword.

INFILE technique

Equivalent to importing the file, you can use the INFILE statement. You set the delimiter and the first observation/row as options for the INFILE statement. You use the INPUT statement to change the variables' names and their data types. Always remember that any INFILE statement is followed by an INPUT one.

Let us try doing that step-by-step. In the Datasets folder, you will find a file with the name Exam.txt. You first upload the file, as we did in Figures 2-21 and 2-22. Get the file path by right-clicking on the file then click Properties, as in Figure 2-34.

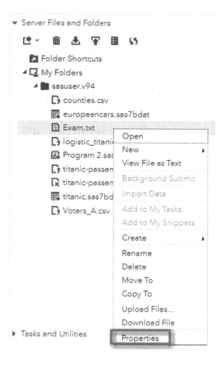

Figure 2-34. *Right-click on the file to get its properties*

Then, select the location and right-click ➤ Copy, as in Figure 2-35.

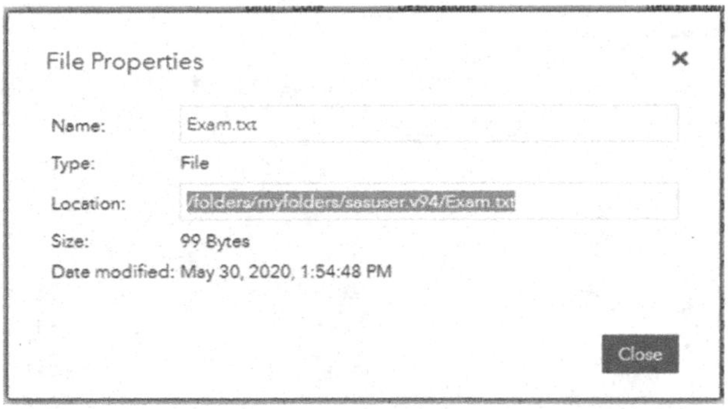

Figure 2-35. *Copy the file location*

The INFILE statement syntax is as follows:

```
INFILE <File Location> DLM=<delimiter> FIRSTOBS=<number of the first
observation/row that you want SAS Studio to read>
```

Listing 2-6 shows an example of using the INFILE statement. The first code line will create a table called examdata saved in the temporary library Work. The second line uses the INFILE statement to specify the raw data file path, uses the option of DLM to indicate the delimiter used, and uses another option of FIRSTOBS to tell SAS Studio to import data rows starting from the second row. The delimiter in this file is a comma. The second row is our first row that contains data. Open examdata.txt and check it. The first row in the file is a header, not data. Let us get back to Listing 2-6. The third line uses the INPUT statement to specify the column names. The names should be separated using spaces. The data step ends with RUN; to tell SAS Studio that this is the last line in this step.

Listing 2-6. INFILE Statement

```
data examdata;
infile '/folders/myfolders/sasuser.v94/Exam.txt' dlm=',' firstobs=2;
input year pass fail;
run;
```

Click Run to execute the program. The output is as in Figure 2-36.

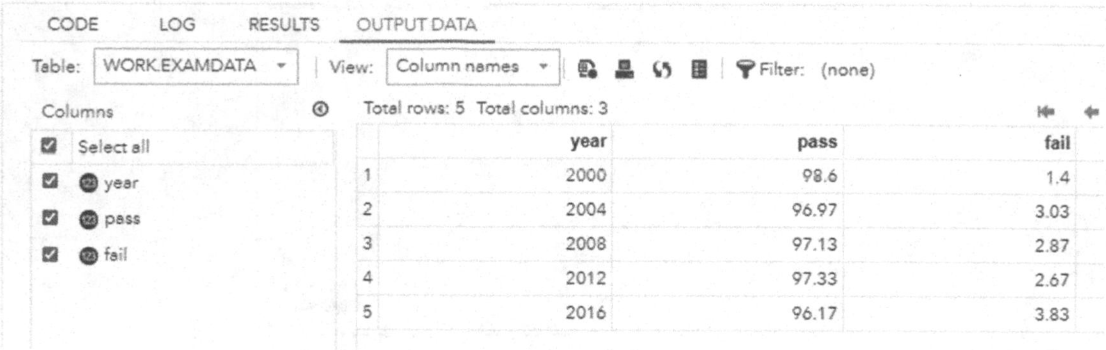

	CODE	LOG	RESULTS	OUTPUT DATA						

Table: WORK.EXAMDATA   ▼ | View: Column names   ▼ | 📇 🖥 ⟳ ▦ | ▼ Filter: (none)

Columns   ⊙   Total rows: 5   Total columns: 3   ⏮ ⏴

☑ Select all		year	pass	fail
☑ ⬤ year	1	2000	98.6	1.4
☑ ⬤ pass	2	2004	96.97	3.03
☑ ⬤ fail	3	2008	97.13	2.87
	4	2012	97.33	2.67
	5	2016	96.17	3.83

Figure 2-36. *The output of the INFILE statement*

Note that FIRSTOBS can be any number, not only zero or one. Assume that you have a big data file of a million rows while you only need to test your code on 100 sample rows. You can set FIRSTOBS=1000 to tell SAS Studio to start reading from row 1000 in that file. To indicate the last row, you use another option to the INFILE statement called OBS. The INFILE statement will be as follows:

```
INFILE <file location> FIRSTOBS=1000 OBS=1100;
```

Summary

This chapter represents the basic foundation of our journey in learning SAS Studio and SAS programming. It starts with the installation steps then explains the interface and the basic components of a SAS program. Also, it explains the data types and the libraries. In the next chapter, we shall dig deeper into the data visualization.

CHAPTER 3

Data Visualization

There is an adage that says, "A picture is worth a thousand words." Visualization is a tedious and harsh task in other programming endeavors. Luckily, SAS Studio can make pretty graphs easily through its integrated development environment (IDE). This chapter uses really big datasets that are available in the public domain. We shall follow the data science process mentioned in Chapter 1, where we start every section with a question and seek its answer through a graph.

This chapter covers various essential charts, such as scatter plots, histograms, series plots, bar charts and sorted bar charts, bubble plots, and cluster analysis and dendrograms. Moreover, this chapter demonstrates how to display data onto maps, which is an extremely challenging task in other programming languages.

Scatter Plot

Scatter plots are used to show the interdependence between variables by describing their direction, strength, and linearity. Moreover, they can easily display the outliers.

For this example, our question is: Do salaries in the city of Seattle depend on the age range of the employees, or is there no relationship?

You can find the data file in the "Datasets" folder of this book's source code called City_of_Seattle_Wages__Comparison_by_Gender_-_Average_Hourly_Wage_by_Age. csv. Or, you can download the CSV file from the Data.Gov project at the following URL: `https://catalog.data.gov/dataset/city-of-seattle-wages-comparison-by-gender-average-hourly-wage-by-age-353b2`.

Copy the CSV file to the "myfolders" folder. From SAS Studio, click Refresh, then import the file as we explained previously in Chapter 2. Change the name of the dataset from IMPORT to Seattle_wages, as in Figure 3-1. Then click Run.

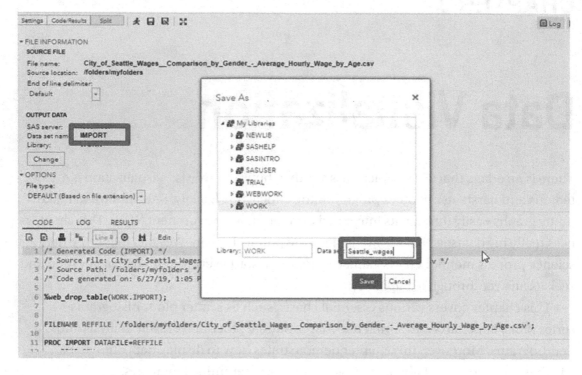

Figure 3-1. *Change dataset name*

To explore the data, click on the Results tab, as in Figure 3-2.

We want to create a scatter plot of the AGE_RANGE and Total_Average_of_HOURLY_ RATE variables to see if there is an interdependency between them. Figure 3-2 shows the two variables.

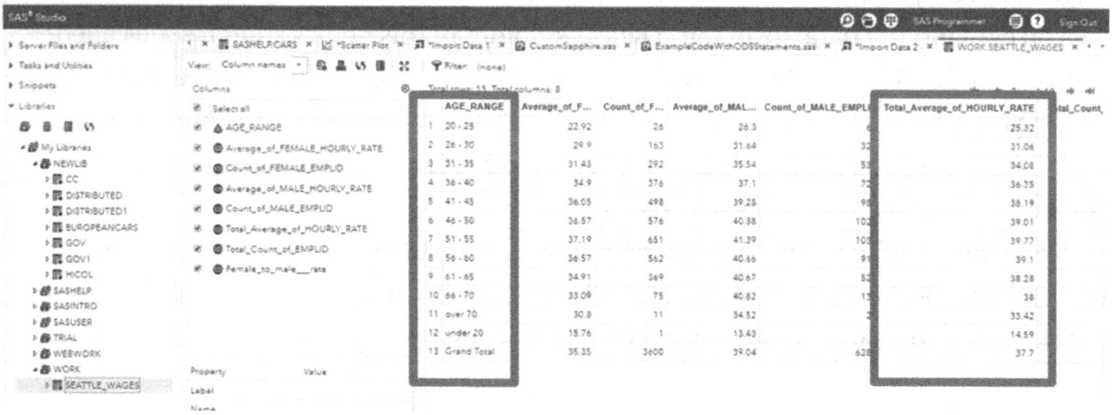

Figure 3-2. *The two variables for the scatter plot*

Click on Tasks and Utilities ➤ Graph ➤ Scatter Plot. In Data, select the WORK. SEATTLE_WAGES dataset. To do that, as in Figure 3-3, click on the small table beside the dataset name.

Figure 3-3. *Click this table to change the dataset*

A new window, as shown in Figure 3-4, will pop up. Select the dataset and click OK.

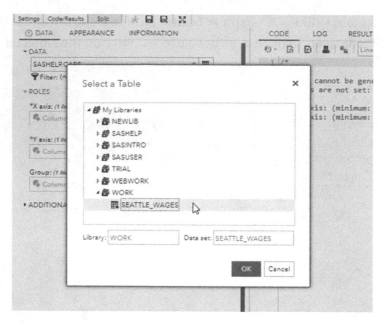

Figure 3-4. *Select the dataset*

Click on the + beside the x-axis and select AGE_RANGE, as in Figure 3-5. Similarly, click on the + beside the y-axis and select Total_Average_of_HOURLY_RATE. Click Run or press F3 to run the program. SAS will auto-generate the code for you in the Code tab, and SAS Studio will display the scatter plot in the Results tab, as in Figure 3-5.

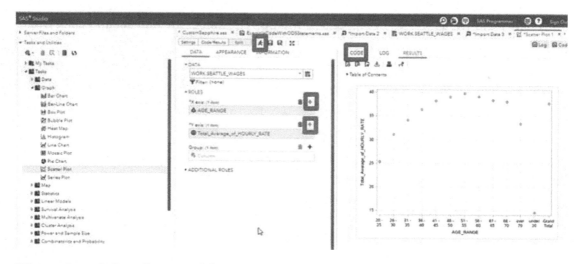

Figure 3-5. *Select the variables and run*

In the graph, the last two values for the AGE_RANGE variable are not needed and give the wrong impression about the relationship between the two variables. Hence, we shall filter and remove them from the data. As in Figure 3-6, click the Filter button under the dataset and type the following:

```
age_range ne 'Grand Total' and age_Range ne 'under 20'
```

Please observe the case letters because while SAS Studio is not case sensitive in the variable names, it is case sensitive in the variables' values. Hence, SAS Studio will interpret "under 20" and "Under 20" differently. It will find nothing for the latter. However, feel free to type age_range as upper- or lowercase or mixed; SAS Studio will interpret it correctly. Now, click Run again to execute the filter code.

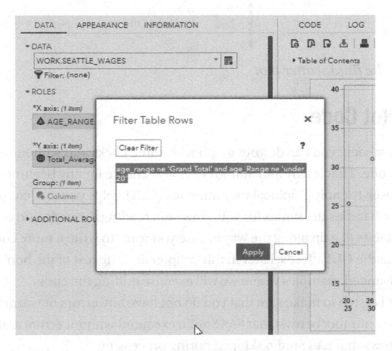

Figure 3-6. *Add the filter code to eliminate the last two values in AGE_RANGE*

From the final scatter plot in Figure 3-7, we can see that there is an interdependence between age and salary in Seattle. The relationship is not linear; it is curvilinear.

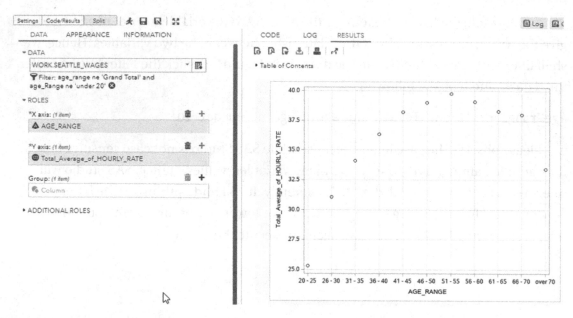

Figure 3-7. *The final scatter plot*

Scatter Plot Code

As you have just seen, you can do your graph with some clicks and without writing a single line of code. That is why SAS Studio is the best choice for any beginner in data science. The user-friendly graphical user interface (GUI) helps you to focus on the concept, and it handles the syntax for you. However, reading the code and checking the log are good habits to gain along the way in case you want to write a more complicated code than what the GUI affords. Later in this chapter and the rest of the book, we shall have some advanced examples where we will exercise editing the code.

Check the Log tab to make sure that you do not have any errors or warnings. Verify from the log the number of rows that SAS Studio executed without errors and resolve the issues of any rows that SAS Studio skipped during processing.

So, let us take a look at the code. The code should look like Listing 3-1.

Listing 3-1. Scatter Plot Code Including Filter Code to Exclude the Last Two Values in AGE_RANGE

```
ods graphics / reset width=6.4in height=4.8in  imagemap;
proc sgplot data= WORK.SEATTLE_WAGES (where=(age_range ne 'Grand Total' and
age_Range ne 'under 20')) ;
scatter x=AGE_RANGE y=Total_Average_of_HOURLY_RATE /
;
    xaxis   grid;
    yaxis   grid;
run;
ods graphics / reset;
```

In the first line of code, SAS Studio uses `ods graphics`. Statistical procedures use ODS Graphics to create graphs as part of their output, setting the width and height of the graph. You can change these dimensions in the code or by using the Appearance tab in the GUI. The second line uses `PROC SGPLOT`, which provides a simple way to make a variety of scatter plots. Notice the filter code is added in the `Data` statement. Then, SAS Studio uses the `scatter` keyword to identify the `sgplot` type and to identify the variables used in the x- and y-axes. Finally, choose xy-grids. You can remove the grids by deleting these two lines from the code or by unchecking them in the Appearance tab.

Scatter Plot Relationships

Figure 3-8 displays various examples of scatter plot relationships. To describe the scatter plot, you should state if it is a linear relationship or curvilinear, as we have seen in the previous example. If the interdependency is linear, look at its direction to know if it is positive or negative. In Figure 3-8, upper-left, the relationship is positive, while the plot at the upper-right is a negative relationship. In Figure 3-8, lower-left, there is no relationship, as the points are scattered irregularly. In Figure 3-8, lower-right, there is no direction, because the relationship is not linear, but rather curvilinear. Finally, scatter plots describe the strength of the relationship. If the points are very close to the line/ form, then there is high interdependency, and the relationship is strong. Otherwise, it is a weak relationship.

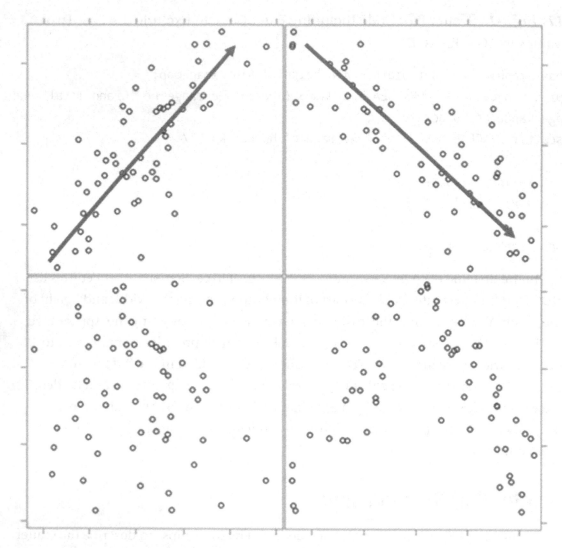

Figure 3-8. *The direction, intensity, and linearity of the scatter plot*

Plotting More Than One Scatter Plot in the Same Image

We can plot more than one scatter plot in the same image to compare values. For this example, we have the number of registered voters in every party and will plot them to compare which party has more registered voters in each of Maine's counties. This program cannot be done using the GUI and has to be done using the code.

The dataset MAINECOUNTIESPARTIES.xlsx in the "Dataset" folder has three columns: the counties and the registered voters per Republican and Democratic party (Table 3-1).

Table 3-1. *MAINECOUNTIESPARTIES.xlsx*

Counties	Republican	Democratic
Androscoggin	28189	22975
Aroostook	19419	13377
Cumberland	57697	102935
Franklin	7900	7001
Hancock	13682	16107
Kennebec	29296	31753
Knox	9148	12440
Lincoln	9727	10241
Oxford	12172	16214
Penobscot	41601	32832
Piscataquis	5403	3098
Sagadahoc	9304	10679
Somerset	14998	9092
Waldo	10378	10442
Washington	9037	6358
York	50388	55828

After uploading the Excel file to SAS Studio and importing it to the Work library, as we did in Chapter 2, we run the code in Listing 3-2.

Listing 3-2. Comparing the Republican Registered Voters to the Democratic Ones in Maine's Counties

```
ods graphics / reset width=6.4in height=4.8in  imagemap;

proc sgplot data= WORK.MAINECOUNTIESPARTIES NOAUTOLEGEND;

scatter x=counties y=Republican markerattrs=(color=red);
scatter x=counties y=Democratic markerattrs=(color=Blue
symbol=circlefilled);
```

```
xaxis grid;
yaxis grid  label="Republican:Red Democrat:Blue";
run;

ods graphics / reset;
```

The code plots two overlapped scatter plots, where each has counties on its x-axis. The y-axis in the first scatter plot is the number of Republicans and in the second one is the number of Democrats. To change the colors to distinguish between each of them, we used the `markerattrs` option. Moreover, we added `symbol=circlefilled` to only the Democratic Party points to distinguish between the points even in grayscale. We added the grids to both the x- and y-axis. Finally, we changed the label of the y-axis to `Republican:Red Democrat:Blue` instead of the legend, as we removed the automatic legend from generation by adding `NOAUTOLEGEND` as an option to `PROC SGPLOT`.

The output of the preceding code is shown in Figure 3-9.

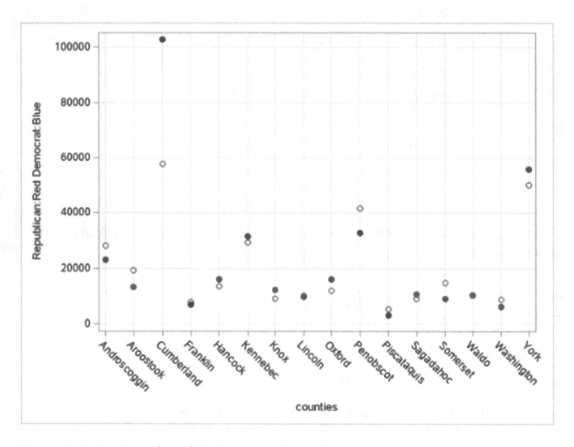

Figure 3-9. *Scatter plot of Maine counties and parties*

From Figure 3-9, we can say that Cumberland County has the highest number of Democratic voters and the highest number of registered voters in general. This might imply that it has the largest population as well, but we cannot ensure that and will need further verification. However, we are sure that the difference between the Democrats and Republicans is enormous there. Other than this county, Democrats are more than the Republicans in eight of the remaining fifteen counties with small differences. Hence, Democrats have to give more care to Maine so it does not turn back to red.

Similarly, you can do the same exercise with your state and get ready for the next presidential election with your own analysis and predictions.

Histogram

The Histogram task creates a chart that displays the frequency distribution of a numeric variable. For this example, we shall see how the annual wages of employees are being distributed in March 2018 in Charlotte, North Carolina.

You will find the data file named City_Employee_Salaries_March_2018.csv in the "Datasets" folder. Or, you can download the CSV file from the Data.Gov project at the following URL: `https://catalog.data.gov/dataset/city-employee-salaries-march-2018`

Copy the CSV file to the "myfolders" folder. From SAS Studio, click Refresh, then import it as we explained in Chapter 2. Change the name of the dataset from IMPORT to Charlotte_wages.

Click on Tasks and Utilities ➤ Graph ➤ Histogram. In Data, select WORK.
CHARLOTTE_WAGES. To do that, as in Figure 3-3, click on the small table beside the
dataset name. Next, click on the + sign beside the Analysis Variable and select Annual_Rt,
as in Figure 3-10; then, click OK.

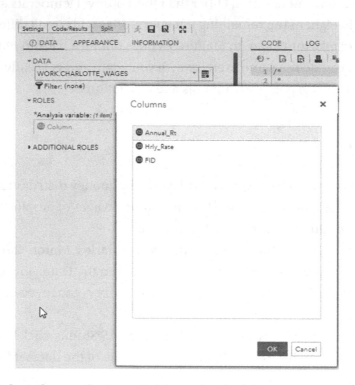

Figure 3-10. *Select the analysis variable to plot its histogram*

Observe that SAS Studio auto-generated the code in the right-hand pane in the Code tab. Again, it uses SG Plot, but using the histogram keyword this time. Now, click Run. SAS Studio will plot the histogram in the Results tab, as in Figure 3-11.

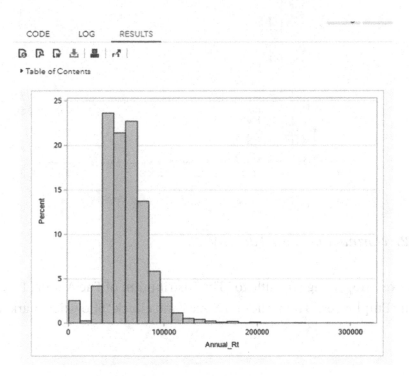

Figure 3-11. *Histogram of Charlotte city employees*

Appearance Tab

Let us enhance the appearance of the graph by comparing our histogram to the normal distribution and kernel estimates. Click on the Appearance tab. Select Normal and Kernel Density. Re-run the code. The output is similar to Figure 3-12.

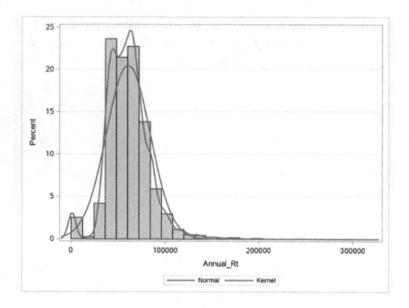

Figure 3-12. *Normal and kernel density*

As in Figure 3-13, change the title to "The Distribution of The Annual Rate of Charlotte City Employees." Then, click on X-Axis and check "Show tick marks at bin midpoints."

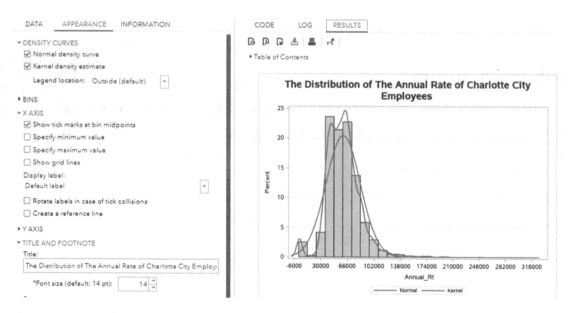

Figure 3-13. *Show tick marks*

Click on the Code tab. Observe how SAS Studio has changed the code according to your choices, as in Figure 3-14.

```
CODE      LOG      RESULTS

⊕▾   ⬚   ⬚   ⬚   ⬚   Line #   ⊙   ⋈   Edit

1
2
3
4
5
6  ods graphics / reset width=6.4in height=4.8in imagemap;
7
8
9  proc sgplot data= WORK.CITY_EMPLOYEE_SALARIES ;
10
11
12
13      title height=14pt "The Distribution of The Annual Rate of Charlotte City Employees";
14
15  histogram Annual_Rt /
16  showbins ;
17
18   density Annual_Rt;
19   density Annual_Rt / type=Kernel;
20
21      yaxis
22              grid       ;
23
24
25
26  run;
27
28  ods graphics / reset;
29   title;
```

Figure 3-14. *Histogram code after changing the settings of the bin midpoints and the title*

Now, return to the Appearance tab. Change the number of bins to 20, as in Figure 3-15. Re-run the code.

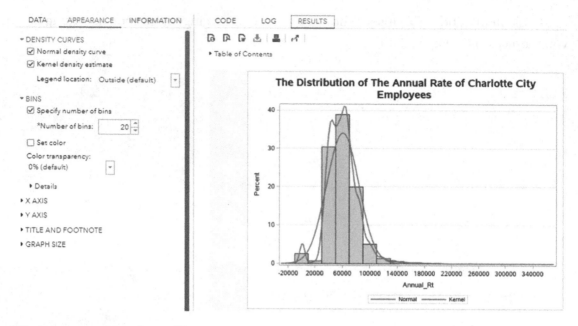

Figure 3-15. *Number of bins*

Now, in the X-Axis section, change the maximum value to 140000, as in Figure 3-16. Re-run.

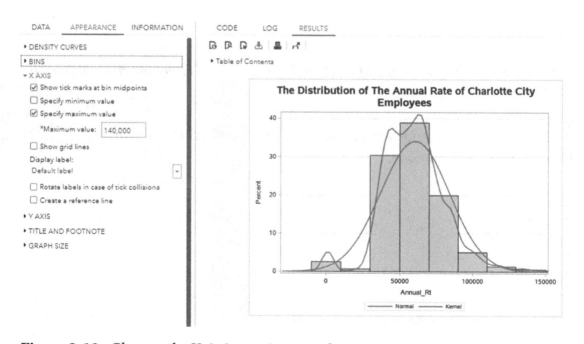

Figure 3-16. *Change the X-Axis maximum value*

Now, we have a more normal distribution curve than before, when we had a right-skewed curve.

Series Plot

The series plot displays the values against time. In this example, let us plot stock trends of prices over time.

Click on Tasks and Utilities ➤ Tasks ➤ Graph ➤ Series Plot. In Data, select SASHELP. STOCKS, for X-axis, select Date, and for Y-axis, select Close, as in Figure 3-17.

Figure 3-17. *Series plot code*

Run the plot. The output is similar to Figure 3-18.

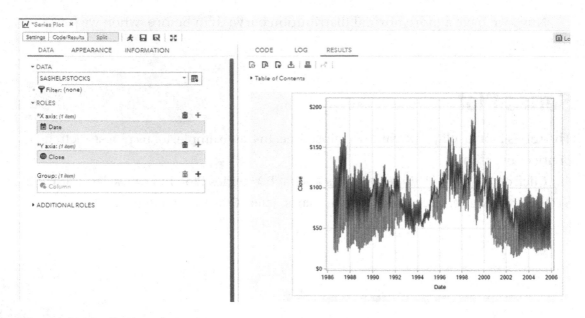

Figure 3-18. *Series plot output*

Let us explore the frequency of the stock and get the unique values of stocks. Click on Tasks and Utilities ➤ Tasks ➤ Statistics ➤ One-Way Frequency. In the Analysis Variable, click on the + sign and select Stock, as in Figure 3-19.

Figure 3-19. *Analysis of the stock using one-way frequency*

You will find that these are stocks for three companies: IBM, Intel, and Microsoft, as in Figure 3-19.

Return again to the series plot and let us enhance the graph upon this information. Since we have three stocks only, select Group By Stock, as in Figure 3-20.

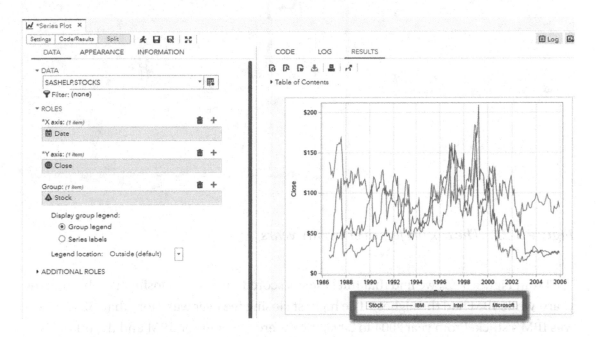

Figure 3-20. *Enhance the series plot by grouping by stock*

The y-axis shows the closing fees, while the x-axis shows the years. Figure 3-21 shows the closing fees for IBM, Intel, and Microsoft from 1986 to 2006.

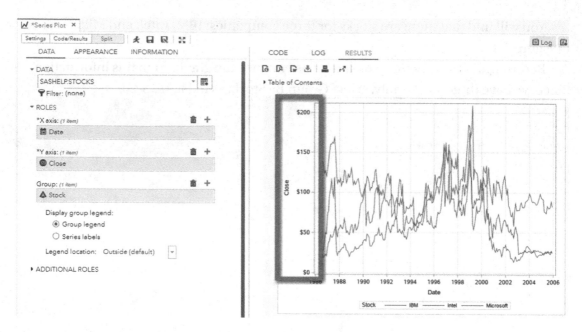

Figure 3-21. *The closing fees over twenty years*

From Figure 3-21, we find that Intel mostly scored the lowest closing fees through the years, while IBM had the highest. The highest closing fee ever was more than $200 and was IBM's stock. From year 2004 to 2006, the difference between IBM and the others is significant, where IBM's stock is close to $100 while Intel and Microsoft are about $25. At that time, this information and data analysis could have helped a stock broker to decide whether to sell or buy which stock.

In the Appearance tab, select Title and Footnote and edit the title to be "Stock Close by Year," as in Figure 3-22. Observe the code changes in the Code tab and re-run the program.

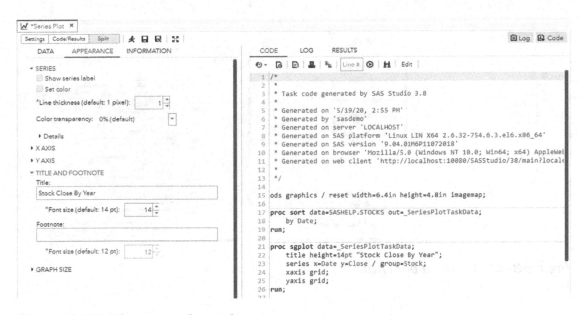

Figure 3-22. *The series plot code*

Bar Chart

You use a bar chart to compare groups or to track a variable over time.

In the following program, you will compare the average salaries of baseball teams. We also want to answer the questions of which team has the highest average salary and which one has the lowest.

Let us make a program to make a horizontal bar chart (HBar) for the average salaries of baseball teams.

Select Tasks ➤ Graph ➤ Bar Chart.

As in Figure 3-23, use the Baseball dataset from the library (SASHELP.BASEBALL). For Chart Orientation, select Horizontal. In Roles, select "Team" for Category, "Salary" for Variable, and "Mean" for Statistic.

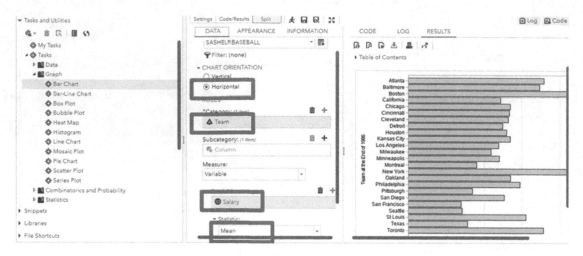

Figure 3-23. *Create a bar chart*

In the Appearance tab, add a report title, "Average Salaries by Team," as in Figure 3-24. Then, click Run.

Figure 3-24. *Add the title*

How Do You Sort a Bar Chart?

Figure 3-24 shows the average salaries by team. However, it can be confusing to find the second and third highest teams from the graph. To avoid any confusion, it is best to sort the graph ascending or descending. We will do sorting the bar chart over four steps.

Step 1: Create a new dataset that has only the team names and mean salaries. Click on Tasks and Utilities ➤ Tasks ➤ Statistics ➤ Summary Statistics. Select SASHELP. BASEBALL for Data, Salary for Analysis Variables, and Team for Classification Variables, as in Figure 3-25.

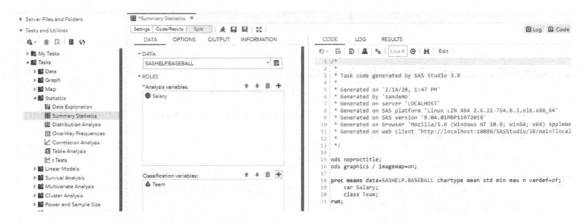

Figure 3-25. *Click on Summary Statistics*

Step 2: In the Options tab, check "Mean" only and deselect the rest of the default selections, as in Figure 3-26.

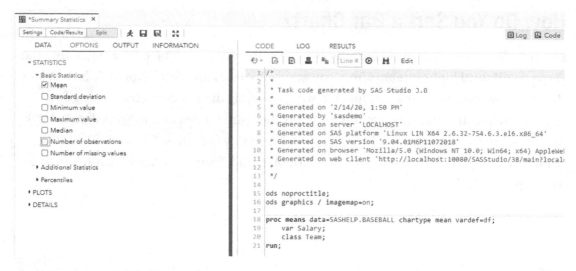

Figure 3-26. *Check "Mean" in the Options tab*

Step 3: In the Output tab, check "Create output data set" to save the mean of the data to a new dataset, as in Figure 3-27.

Figure 3-27. *Save the results in a new dataset*

Step 4: Make a bar chart sorted by the mean salary, not by the team. Select the output dataset that was generated from the previous step as the input data to a new bar chart. Select Salary_Mean for Category, Team for Subcategory, and Salary_Mean for the Measure Variable, as in Figure 3-28.

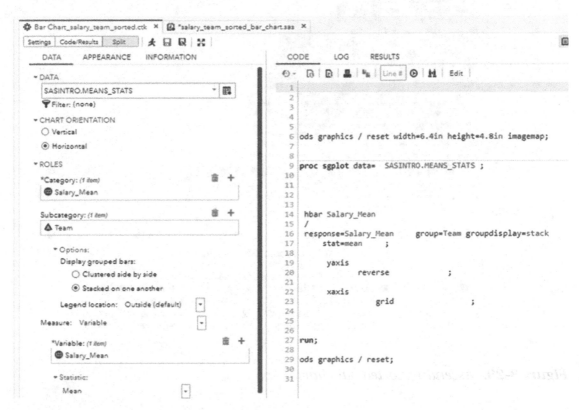

Figure 3-28. *Select your table as your data*

Run the chart to get a new sorted graph, as in Figure 3-29.

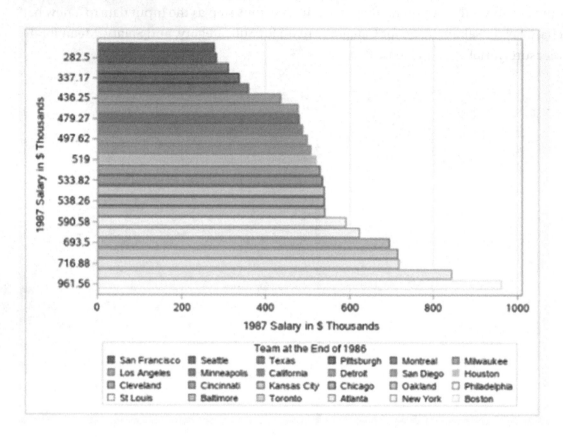

Figure 3-29. *Ascending sorted bar chart*

To reverse the output, check "Reverse tick values" in Category Axis in the Appearance tab, as in Figure 3-30.

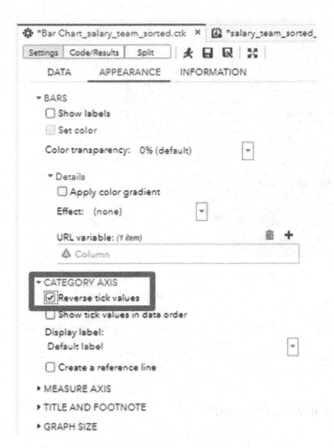

Figure 3-30. *Check "Reverse Tick Values"*

Re-run the bar chart. The output is similar to Figure 3-31.

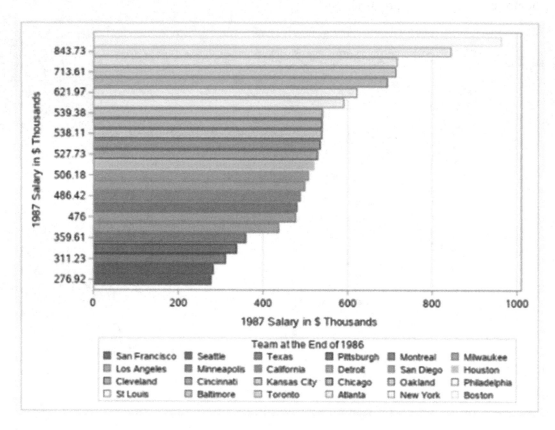

Figure 3-31. *Descending bar chart*

Finally, the team that has the highest average salary in 1987 is Boston, and the one that has the lowest average salary is San Francisco.

Create a Histogram Using a Bar Chart

A bar chart can be used to create histograms as well. In this example, you will learn how to create a chart of the presidential candidates and their winning percentages in Maine in 2012, which was previously mentioned in Chapter 1 in Figure 1-9. The results in Maine were as shown in Table 3-2.

Table 3-2. *Results of Candidates and Their Winning Percentages in Maine in 2012*

Candidates	Winning Percentages
Obama	56.27
Romney	40.98
Johnson	1.31
Stein	1.14
Paul	0.29
Anderson	0.01
Reed	0

You will find a dataset created to present these percentages in the "Dataset" folder; the file is named Maine_Candidates and their winning percentages in 2012.csv.

Copy the CSV file to your "myfolders" folder. From SAS Studio, click Refresh, and then import it as we explained in Chapter 2. Change the name of the dataset from IMPORT to CANDIDATES2012, as in Figure 3-32. Click Run.

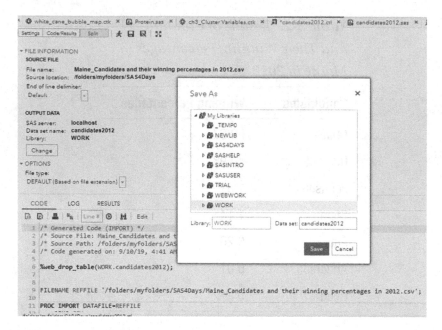

Figure 3-32. *Import the dataset*

To create a histogram using a bar chart, click on Tasks and Utilities ➤ Tasks ➤ Graph ➤ Bar Chart. As in Figure 3-33, in the Data section, select the WORK library and CANDIDATES2012 table. In Roles ➤ Category, select the column x. Then, run the program.

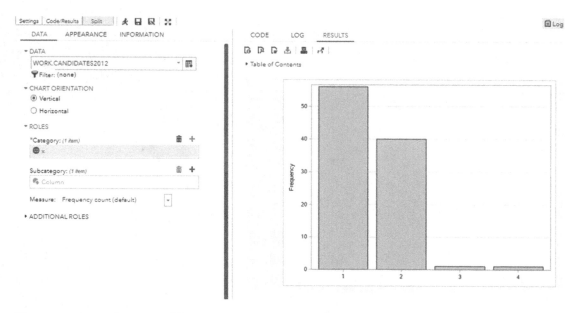

Figure 3-33. *The initial histogram using bar chart*

Now, we want to do three tasks to enhance the chart:

- Color the bars with the corresponding party color.

- Modify the footnote to append the candidates' names.

- Remove the x-axis label.

To do these three tasks, we need to edit the code, and not from the GUI. In the Code tab, click Edit. The code will open in a new program window. Edit the code to be as in Listing 3-2, or just copy and paste the listing, which is in the source code for this book.

Listing 3-3. Histogram Using Parties' Colors

```
ods graphics / reset width=6.4in height=4.8in imagemap ;

proc sgplot data=  WORK.CANDIDATES2012 noautolegend;

        footnote2 "The candidates, 1: Obama 2:Romney 3:Johnson, 4:Stein
        5:Paul 6:Anderson 7:Reed";

        vbar x / colorresponse=x colorstat=mean colormodel=(blue red silver
        green) ;
        xaxis display=(nolabel);
        yaxis grid;
run;
ods graphics / reset;
footnote2;
```

Run the code. The output will be as in Figure 3-34.

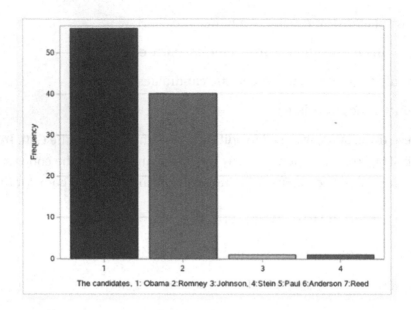

Figure 3-34. *The bars with corresponding party colors*

Please feel free to play with the code. For example, remove `noautolegend` in the second line of the code, re-run it, and check the change in the graph; or remove `display=(nolabel)` in the X-Axis options or play with the colors in the colormodel option.

For more examples and explanations of the bar chart options, please check the following links:

1. SAS Documentation—Managing Axes in OVERLAY Layouts:
   `https://documentation.sas.com/?docsetId=grstatug&docset Target=p1pqfzgbuzbpkzn1mrbzhgggvhkz.htm&docsetVersion=9. 4&locale=en#p18of8gjhjfti1n15y1w0iqtp4nr`

2. SAS Blog—Response Colors and Thickness:
   `https://blogs.sas.com/content/graphicallyspeaking/ 2015/09/21/response-colors-and-thickness/`

3. Data Flair—Types of Bar Chart in SAS:
   `https://data-flair.training/blogs/sas-bar-chart/`

Bubble Chart

The scatter plot is useful if you want to display the relationship between two variables. If you want to display the relationship between three variables, you use the bubble chart. In this example, we will create a bubble chart to display the heights, weights, and ages of students of a class.

Click on Tasks and Utilities ➤ Tasks ➤ Graph ➤ Bubble Chart. In Data, select the SASHELP.CLASS library. Then in Roles, for the X axis of the graph, select Height; for the Y axis, select Weight, and for Bubble size, select Age, as in Figure 3-35.

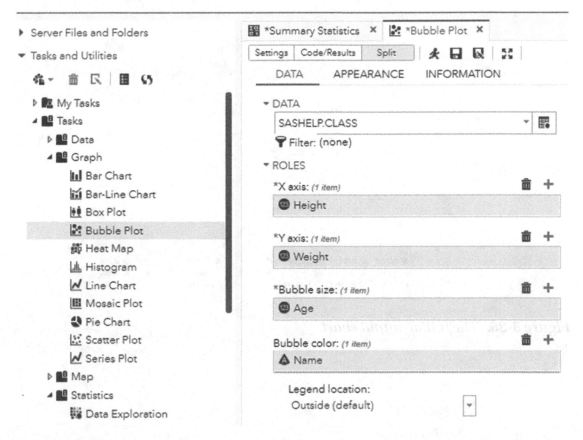

Figure 3-35. *Bubble chart*

Click Run. The initial output graph will be as in Figure 3-36.

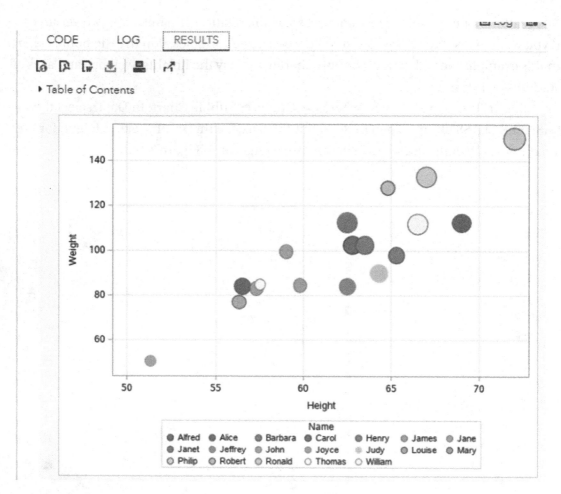

Figure 3-36. *The initial output chart*

Now, let us enhance the appearance. Click on the Appearance tab. As in Figure 3-37, in Bubble Labels, for Bubble Label, select Name. As in Figure 3-37, click on Title and Footnote, and for the Title, write "Age by Height and Weight," and for Footnote, write "Bubble size represents age."

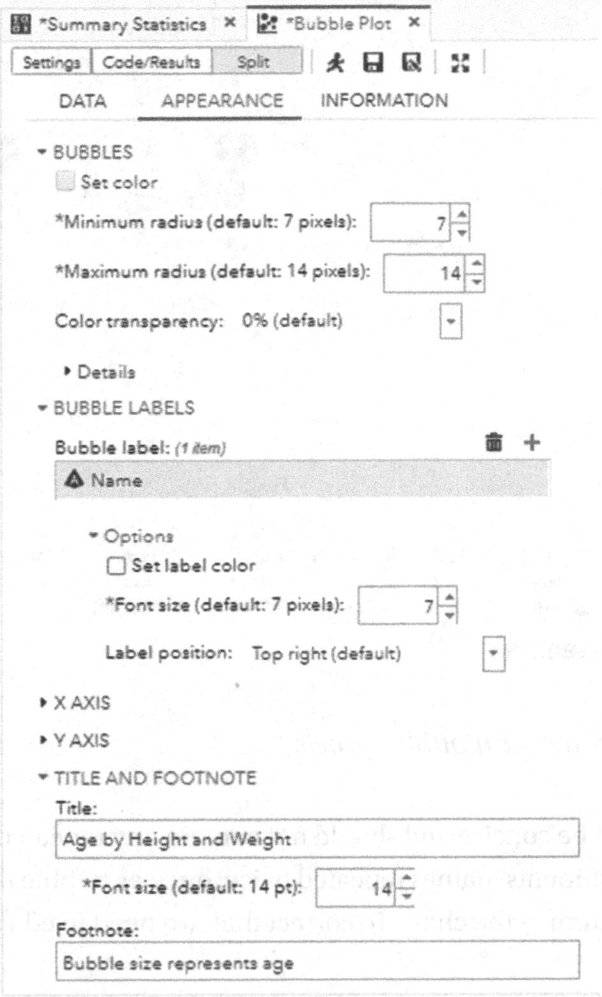

Figure 3-37. *Add bubble label*

Re-run the program. The output will be as in Figure 3-38.

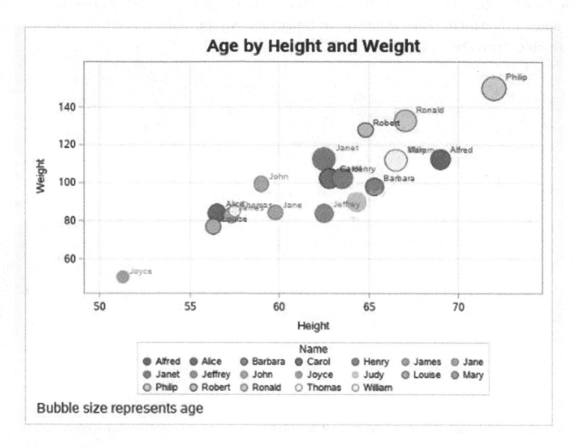

Figure 3-38. *The chart with bubble labels*

The chart should be concise and should not contain any repeated data. The graph in Figure 3-38 has the students' names repeated twice, once as bubble labels and again in the legend at the bottom of the chart. To correct that, we need to edit the code. Click Edit, as in Figure 3-39.

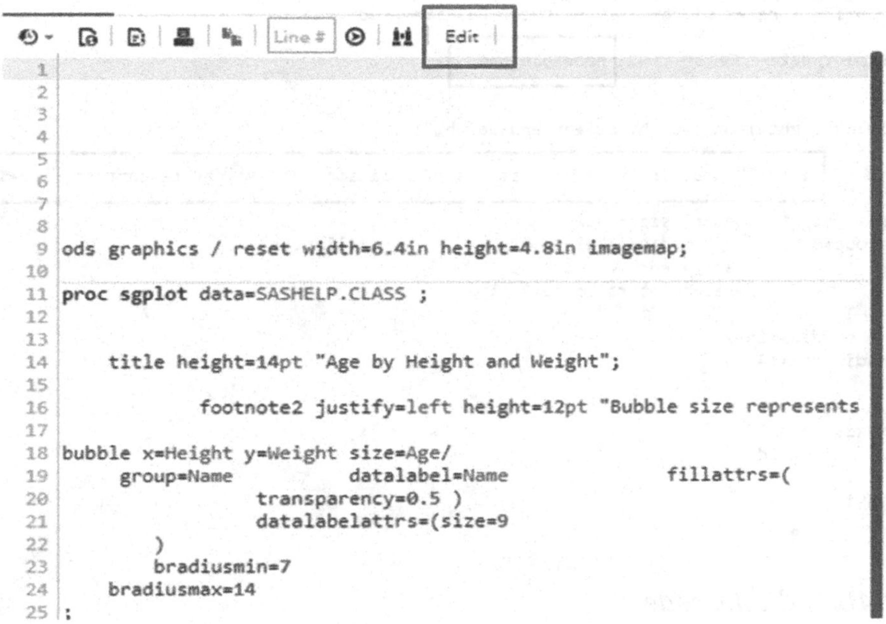

```
🕘 ▾   ⬜   ⬜   ⬛   ⬛   Line #   ⊙   ⬛   Edit
 1
 2
 3
 4
 5
 6
 7
 8
 9  ods graphics / reset width=6.4in height=4.8in imagemap;
10
11  proc sgplot data=SASHELP.CLASS ;
12
13
14      title height=14pt "Age by Height and Weight";
15
16              footnote2 justify=left height=12pt "Bubble size represents
17
18  bubble x=Height y=Weight size=Age/
19      group=Name            datalabel=Name            fillattrs=(
20                  transparency=0.5 )
21                  datalabelattrs=(size=9
22          )
23          bradiusmin=7
24      bradiusmax=14
25  :
```

Figure 3-39.   *Click Edit*

As in Figure 3-40, edit the code by adding the option of noautolegend to the sgplot function to remove the auto-generated legend under the graph. To enhance the appearance of the footnote, let us change its place to be inside the graph and indent it to the bottom right so as to not overlap any of the results. To do that, add the following line of code:

```
inset"Bubble size represents Age" / position=bottomright
textattrs=(size=11);
```

```
ods graphics / reset width=6.4in height=4.8in imagemap;

proc sgplot data=SASHELP.CLASS noautolegend;

    title height=14pt "Age by Height and Weight";
        inset "Bubble size represents Age" / position=bottomright textattrs=(size=11);

bubble x=Height y=Weight size=Age/
    group=Name              datalabel=Name                      fillattrs=(
                transparency=0.5 )
                datalabelattrs=(size=9
        )
        bradiusmin=7
    bradiusmax=14
;

    xaxis
            grid                    ;

    yaxis
            grid            ;
```

Figure 3-40. *Edit the code*

Now, re-run the code. The output is enhanced and would be as in Figure 3-41.

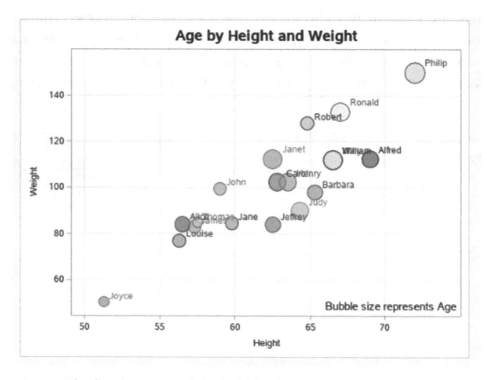

Figure 3-41. *The final version of the bubble chart*

From this plot, we can see that Joyce is the lightest and shortest student in the class. Also, she is one of the two youngest students in the class. Joyce and Thomas have the smallest circles. If we return to the table by clicking on Libraries ➤ SASHELP ➤ CLASS, click on the Age column to sort the data by age. You will find that both of them are eleven years old. On the other hand, Philip is the tallest, heaviest, and oldest student in the class. He is the only student who is sixteen years old, and the rest of the class is younger than him. That is why his circle is the largest. Joyce and Philip would be considered as outliers. However, this class is only nineteen students, which is a very small sample. The larger the sample size, the better the analysis.

Maps

Displaying data on a map adds new perspectives. It can tell a whole story all by itself. SAS Studio made displaying data on maps easy. Moreover, customizing the map with colors and bubbles adds new dimensions to your data. This section will walk you through the process using real-life data. It will show you in step-by-step how to download the data and prepare it to be displayed on a map.

Bubble Map

The Bubble Map task creates a map that is overlaid with a bubble plot. For this example, we shall use crime data for New York City to create a bubble map showing the crime rate in the different NYC boroughs in 2018. The data is available through the NYC Open Data Project website.

You will find the data file in the "Datasets" folder; the file is named NYPD_Complaint_Data_Current__Year_To_Date.csv. Or, you can download the dataset from this URL: https://data.cityofnewyork.us/Public-Safety/NYPD-Complaint-Data-Current-Year-To-Date-/5uac-w243/data

To download the dataset from the link, as in Figure 3-42, click on Export, then on CSV.

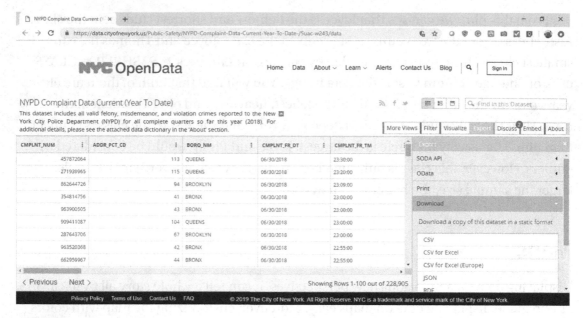

Figure 3-42. *NYPD Compliant Data—NYC OpenData*

The dataset contains a massive amount of information. Much investigation and analysis can be done with this data. For example, what is the most prevalent crime type in every borough? When is the most probable time for crime to happen? On which day of the week do most crimes occur?

For this example, we are interested solely in answering the question of which borough has the highest crime rate and which has the least.

To answer this question, we need the following three columns: Borough, Longitude, and Latitude. Hence, we drop all the other columns. Then, we count the number of crimes in every borough to represent it with a bubble on a map.

After cleaning the data and preparing it, we have the dataset shown in Table 3-3.

Table 3-3. *Number of Crimes Per Borough*

Borough	Count of Crimes	Latitude	Longitude
BRONX	50153	40.82452	-73.8978
BROOKLYN	67489	40.82166	-73.9189
MANHATTAN	56691	40.71332	-73.9829
QUEENS	44137	40.74701	-73.7936
STATEN ISLAND	10285	40.63204	-74.1222

To load the data, we write Listing 3-4 in SAS Studio. You can load Listing 3-4. sas from the "Example Code" folder. Click Run.

Listing 3-4. Creating a Dataset for Number of Crimes Per Borough

```
data NYC_crime;
input Borough $13.;
datalines;
BRONX
BROOKLYN
MANHATTAN
QUEENS
STATEN ISLAND
;
run;
data NYC_crime_dim;
set nyc_crime;
input count Latitude Longitude;
datalines;
50153 40.82451851 -73.897849
67489 40.82166423 -73.91885603
56691 40.71332365 -73.98288902
44137 40.74701026 -73.79358825
10285 40.63203653 -74.1222402
;
run;
```

In this program, we did not import the raw data file. We inserted the values of the table into the code using the DATALINES keyword in the data step. In Listing 3-4, we have two data steps. The first one is called NYC_crime. The second line uses the INPUT statement to specify the columns' names and types. In this table, we define only one column as the borough names, and its type is character because we used the $. The maximum number of characters is 13. Then, we use a DATALINES statement to insert the actual values, and end it with a semicolon.

The second data step creates another table called NYC_crime_dim to insert the boroughs' dimensions and the number of crimes in each one. This table is initialized by the previous table, nyc_crime, by using the SET statement. Again, we use the INPUT statement to specify that we shall add three more columns to the borough. We add the count, latitude, and longitude. Remember to leave a space in between the column names, and do not use commas. Again, use DATALINES to insert the actual values, and end it with a semicolon.

After running Listing 3-4, two new tables called NYC_crime and NYC_crime_dim will be created in the Work library. To check the output table, click Libraries ➤ Work; under it, you will find NYC_crime and NYC_crime_dim. If not, refresh your libraries as we explained in Chapter 2.

Now, click on Tasks and Utilities ➤ Tasks ➤ Map ➤ Bubble Map, as in Figure 3-43.

Figure 3-43. *Create a bubble map*

As in Figure 3-44, select the dataset WORK.NYC_CRIME_DIM in the Data field. In Roles, select the Latitude and Longitude. In Bubble size, select count, which is the number of crimes. Finally, in Group, select the Borough column.

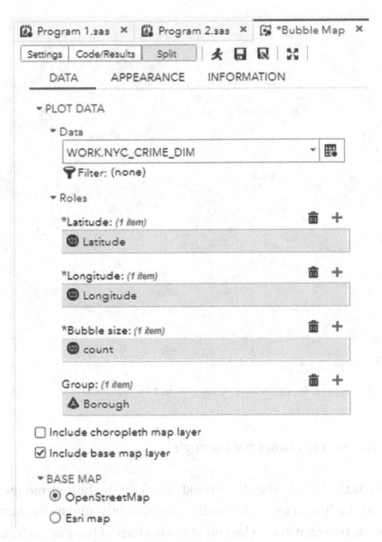

Figure 3-44. Number of crimes per borough

For Base Map select OpenStreetMap. Later in the chapter, we shall try the Esri maps to see the difference in the output. Leave the rest of defaults and click Run. The bubble map seen in Figure 3-45 will be displayed.

CODE   LOG   RESULTS

▸ Table of Contents

Figure 3-45. *Number of crimes per borough*

It is clear from the bubble size that Brooklyn has the highest crime rate, while Staten Island has the lowest one. SAS Studio automatically adjusts the bubble location, size, and color. Moreover, it adds a legend at the bottom of the map for the borough color codes.

Now, let us enhance the appearance of the bubble map. We can add a label to the bubble with the count of crimes, and add a title to the map. As in Figure 3-46, in the Appearance tab, in Data Labels, for Bubble label, select the count column from the dataset. Then, in Label options, for Font weight, select Bold. In the Label position, select Center. Finally, in Title And Footnote, for Title, type "Number of crimes in NYC Boroughs."

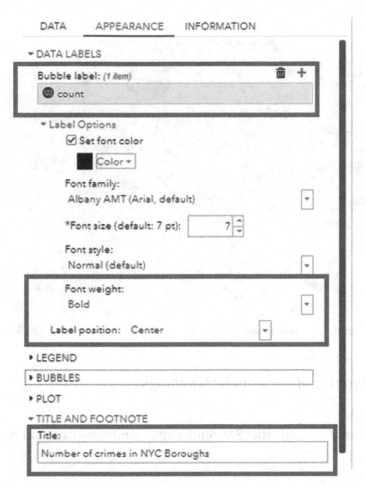

Figure 3-46. *Enhance the appearance*

Click Run to check the new changes. The output would be as in Figure 3-47.

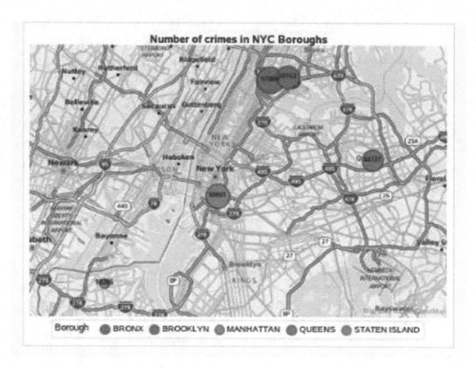

Figure 3-47. *Number of crimes in NYC boroughs*

Listing 3-5 shows the code that SAS Studio auto-generates from our options through the user interface.

Listing 3-5. Creating a Dataset for Number of Crimes Per Borough

```
ods graphics / reset width=6.4in height=4.8in;
proc sgmap plotdata=WORK.NYC_CRIME_DIM;
openstreetmap;
title 'Number of crimes in NYC Boroughs';
bubble x=Longitude y=Latitude size=count/    group=Borough datalabel=count
datalabelpos=center
datalabelattrs = (color=CX0f0e0e  size=7pt  weight=bold)
name="bubblePlot";
    keylegend "bubblePlot" / title =  'Borough' ;
run;
ods graphics / reset;
 title;
```

SAS Studio is powerful in that it provides two types of base maps: OpenStreetMap and Esri maps. Now, let us try the other base map type. As in Figure 3-48, return to the Data tab, choose Esri maps, and click Run. The output map will be as in Figure 3-48.

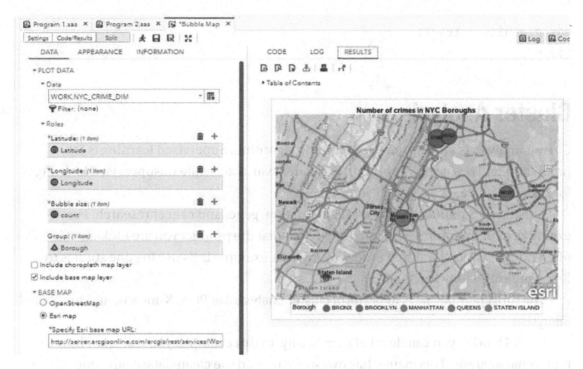

Figure 3-48.   *The map using Esri maps*

SAS Studio will generate the code in Listing 3-6 for this Esri map option.

Listing 3-6.   Generate the map Using Esri Map

```
ods graphics / reset width=6.4in height=4.8in;

proc sgmap plotdata=WORK.NYC_CRIME_DIM;

        esrimap url= 'http://server.arcgisonline.com/arcgis/rest/services/
                World_Street_Map/MapServer';

    title 'Number of crimes in NYC Boroughs';

    bubble x=Longitude y=Latitude size=count / group=Borough
        datalabel=count datalabelpos=center
```

```
            datalabelattrs=(color=CX0f0e0e size=7pt weight=bold)
            name="bubblePlot";
    keylegend "bubblePlot" / title ='Borough';
run;

ods graphics / reset;
title;
```

Cluster Analysis

Cluster analysis is an unsupervised learning algorithm. Supervised learning is for prediction and has a specific output/dependent variable, while unsupervised learning is for exploring data.

Clustering is popular in market segmentation, gene, and cancer research. For example, in online stores, you usually will find that the product you are looking at is often bought with other listed products. Another example is its use in gene studies to find similarities between them.

The most popular ways to perform cluster analysis are PCA, K-means, and hierarchal analysis.

In SAS Studio, you can do all of them easily. In this example, you will learn how to do hierarchal analysis. This mainly has two steps: upload the clean dataset and add all the variables to the cluster, and SAS Studio will handle the rest for you.

Click on Tasks and Utilities ➤ Tasks ➤ Cluster Analysis ➤ Cluster Variables. The user interface for the Cluster Variables task opens. As in Figure 3-49, on the Data tab, select the SASHELP.CARS dataset. Finally, select all the variables you want in the Variables to cluster section. Then, click Run.

Figure 3-49. *Cluster Variables*

Hierarchal clustering has two types: agglomerative (bottom-up) and divisive (top-down).

The agglomerative algorithm considers each point as a cluster, then repeatedly combines the two nearest clusters. The divisive starts with one cluster and recursively splits it into more branches.

SAS Studio uses the divisive method. The output clearly explains how the tree is structured and then how each branch is divided into more branches. In this example, the output, which is shown in Figure 3-50, says that the cluster is divided into two branches, and that on the second level, each branch is divided into more branches.

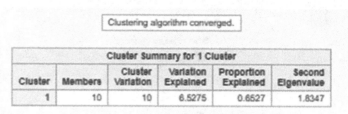

Clustering algorithm converged.

		Cluster Summary for 1 Cluster			
Cluster	Members	Cluster Variation	Variation Explained	Proportion Explained	Second Eigenvalue
1	10	10	6.5275	0.6527	1.8347

Total variation explained = 6.5275 Proportion = 0.6527

Cluster 1 will be split because it has the largest second eigenvalue, 1.834683, which is greater than the MAXEIGEN=1 value.

Clustering algorithm converged.

		Cluster Summary for 2 Clusters			
Cluster	Members	Cluster Variation	Variation Explained	Proportion Explained	Second Eigenvalue
1	7	7	5.144403	0.7349	0.9396
2	3	3	2.769639	0.9232	0.2295

Total variation explained = 7.914042 Proportion = 0.7914

2 Clusters		R-squared with			
Cluster	Variable	Own Cluster	Next Closest	1-R**2 Ratio	Variable Label
Cluster 1	Engine Size	0.8291	0.4443	0.3075	Engine Size (L)
	Cylinders	0.7326	0.5291	0.5679	
	MPG_City	0.7284	0.3136	0.3957	MPG (City)
	MPG_Highway	0.7404	0.2755	0.3583	MPG (Highway)
	Weight	0.8570	0.2753	0.1973	Weight (LBS)
	Wheelbase	0.6474	0.0550	0.3732	Wheelbase (IN)
	Length	0.6095	0.0605	0.4156	Length (IN)
Cluster 2	MSRP	0.9658	0.2426	0.0451	
	Invoice	0.9637	0.2366	0.0475	
	Horsepower	0.8401	0.5302	0.3403	

Figure 3-50. *Splitting the cluster*

Moreover, SAS Studio publishes dendrograms, as in Figure 3-51. To read the dendrogram in the order of the cluster formation, read it from right to left.

However, to know the similarity, read the dendrogram from left to right. The cluster of Horsepower, Invoice, and MSRP is most similar as it has the shortest branch. Similarly, the other cluster of Length, Wheelbase, Weight, MPG (Highway), MPG (City), Cylinders, and Engine Size is similar and has the shortest branch equal to that of the previous cluster. The first cluster has the longest branch.

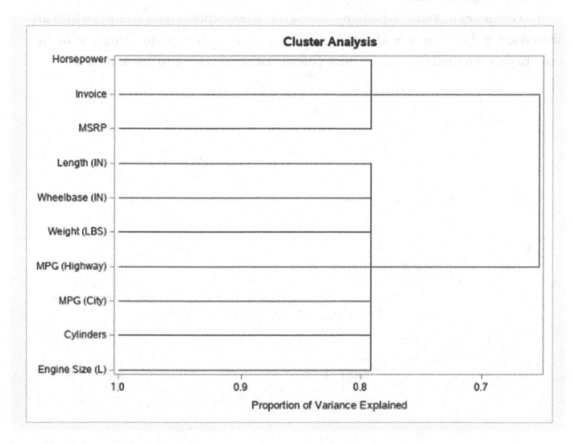

Figure 3-51. *Dendrogram*

Summary

This chapter digs deeper into data visualization. It starts with the most essential plots, such as scatter plots and histograms, then moves to more advanced ones, such as maps. Moreover, it shows how to concatenate plots over each other by changing the colors and patterns to ease the comparisons of the findings. Data visualization in SAS Studio is one of the most powerful features because plenty of the plots can be done using the IDE and without writing any code.

The chapter ends with explaining some crucial concepts in data science, which are supervised and unsupervised learning. Also, it talks about how to do clustering. In the next chapter, we shall explain the most crucial statistical tasks and linear models.

PART II

More Programming

Statistical Analysis and Linear Models

In Chapter 3, we explored the graph and map tasks. In Chapter 4, we will explore statistics and linear models.

Statistical Analysis

In Tasks ➤ Statistics tab, you will find some of the statistical tasks that should be included in the reports. In this chapter, we will discuss One-Way Frequency, Summary Statistics, Correlation Analysis, and T-Tests.

One-Way Frequency

The One-Way Frequencies task is usually the first step when exploring any dataset. It generates frequency tables from your data showing the unique levels of each variable. You can also use this task to perform binomial and chi-square tests. You can use this task to distinguish the outliers as well.

For this example, we explore the SASHELP.HEART dataset to see how many people have high cholesterol. Click on Tasks and Utilities ➤ Tasks ➤ Statistics ➤ One-Way Frequencies. In Data, select SASHELP.HEART, and in Analysis Variables, select CHOL_STATUS, as in Figure 4-1.

© Engy Fouda 2020
E. Fouda, *Learn Data Science Using SAS Studio*, https://doi.org/10.1007/978-1-4842-6237-5_4

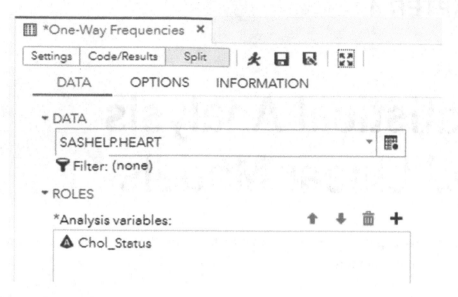

Figure 4-1. *Choose dataset and analysis variable*

In the Options tab, select Frequencies and Percentages, and then check "Frequency table" and "Include percentages," and in Row value order, select Descending frequency, as in Figure 4-2.

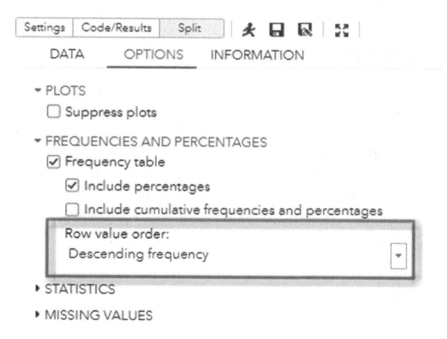

Figure 4-2. *Change the options*

Then, click Run. The output is a frequency table, as shown in Figure 4-3, and a graph, as shown in Figure 4-4.

Cholesterol Status		
Chol_Status	Frequency	Percent
Borderline	1861	36.80
High	1791	35.42
Desirable	1405	27.78
Frequency Missing = 152		

Figure 4-3. *Frequency table*

The table in Figure 4-3 shows that Chol_Status has only three unique levels: Borderline, High, and Desirable. It shows the frequency of each one and its percentage of appearance in the dataset.

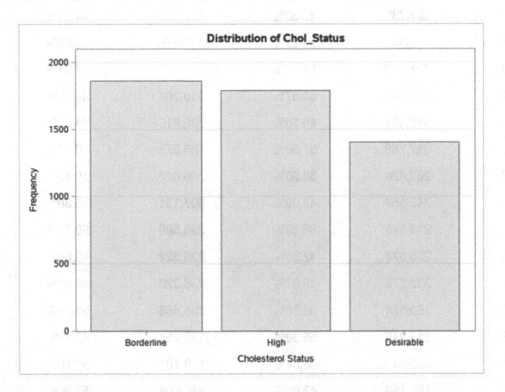

Figure 4-4. *Distribution of Chol_Status*

Figure 4-4 shows the distribution of Chol_Status in descending order, where Borderline has the highest percentage in this dataset, followed by High, and the lowest frequency is Desirable.

Summary Statistics

The Summary Statistics task provides descriptive statistics for variables across all observations and within groups of observations. You can also summarize your data in a graphical display, such as a histogram or box plot.

For this example, we review Maine's past elections to see if there are any predictable outcomes (patterns). The dataset is in the "Dataset" folder and has the name Maine_Past_Elections.xlsx. From SAS Studio, upload it and click Refresh, then import it as we explained in Chapter 2. Change the name of the dataset from IMPORT to PAST_ELECTIONS. The contents of the Excel file are as shown in Table 4-1.

Table 4-1. *Maine_Past_Elections.xlsx*

Year	D	%	R	%
2016	**357,735**	**47.80%**	335,593	44.90%
2012	**401,306**	**56.27%**	292,276	40.98%
2008	**421,923**	**57.71%**	295,273	40.38%
2004	**396,842**	**53.57%**	330,201	44.58%
2000	**319,951**	**49.10%**	286,616	44.00%
1996	**312,788**	**51.60%**	186,378	30.80%
1992	**263,420**	**38.80%**	206,820	30.40%
1988	243,569	43.90%	**307,131**	**55.30%**
1984	214,515	38.80%	**336,500**	**60.80%**
1980	220,974	42.30%	**238,522**	**45.60%**
1976	232,279	48.07%	**236,320**	**48.91%**
1972	160,584	38.50%	**256,458**	**61.50%**
1968	**217,312**	**55.30%**	169,254	43.10%
1964	**62,264**	**68.84%**	118,701	31.16%
1960	**181,159**	**42.95%**	240,608	57.05%

Click on Tasks and Utilities ➤ Tasks ➤ Statistics ➤ Summary Statistics. In Data, select the WORK.PAST_ELECTIONS table, and in Roles select "D" and "R" as the Analysis variables, as in Figure 4-5.

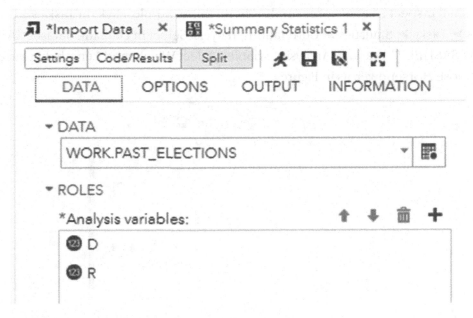

Figure 4-5. *Summary Statistics Data tab*

Click Run. The output will be as shown in Figure 4-6.

Variable	Label	Mean	Std Dev	Minimum	Maximum	N
D	D	267108.07	100623.20	62264.00	421923.00	15
R	R	255776.73	65083.49	118701.00	336500.00	15

Figure 4-6. *Summary statistics of Democrats and Republicans in Maine*

The difference between the Democrats' and Republicans' means is not remarkable. Therefore, Maine is "Lean Democrat," as Democrats won seven elections out of fifteen (from the year 1960 to 2016). Since 1992, the Democratic Party has always won the elections in Maine.

Correlation Analysis

Correlation is a statistical procedure for describing the relationship between numeric variables. The relationship is described by calculating correlation coefficients for the variables. The correlations range from –1 to 1. The Correlation Analysis task provides graphs and statistics for investigating associations among variables.

We shall inspect a couple of examples. To do this example, click on Tasks and Utilities ➤ Tasks ➤ Statistics ➤ Correlation Analysis. Select in Data the dataset CARS from the SASHELP library. In Analysis variables, select Weight and Length. In Correlate with, choose Horsepower, as in Figure 4-7.

Figure 4-7. *Correlation Analysis Data tab*

In Options ➤ Plots, select Individual Scatter Plots, as in Figure 4-8.

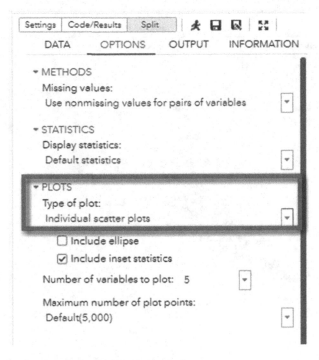

Figure 4-8. *Correlation Analysis Options tab*

Then, click Run. The output will be as shown in Figure 4-9.

1 With Variables:	Horsepower
2 Variables:	Weight Length

Pearson Correlation Coefficients, N = 428		
	Weight	Length
Horsepower	0.63080	0.38155

Figure 4-9. *Pearson correlation coefficients*

The output shows that the horsepower is more correlated with weight than length. Again, these correlations are confirmed in the following plots, Figures 4-10 and 4-11.

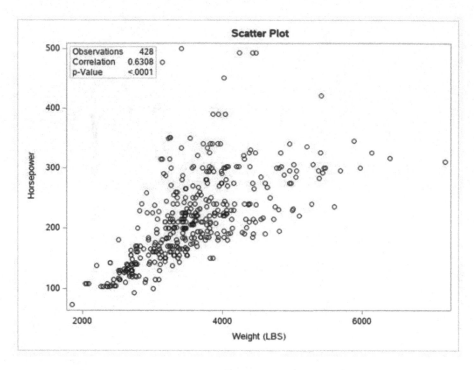

Figure 4-10. *Correlation between horsepower and weight*

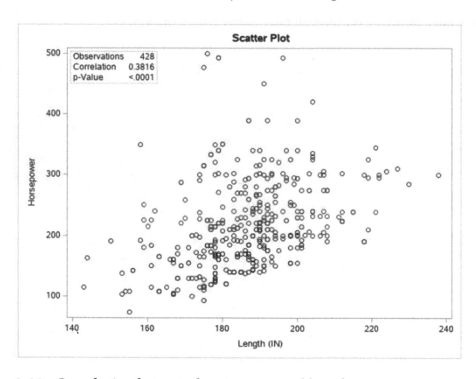

Figure 4-11. *Correlation between horsepower and length*

Now, let us look at how to use correlation for the presidential election in Maine. The question is: Does the gross domestic product (GDP) affect the voters turnout in the state of Maine?

The dataset is called correlation.xlsx and can be found in the "Datasets" folder. I faced a challenge while building this dataset. Various media and government sites put out different values for Maine's voter turnout. I used the most repeated numbers from 2016. Further verification is needed if you will use these numbers beyond this exercise.

Here is a sample of the references that I researched:

- https://en.wikipedia.org/wiki/United_States_presidential_election_in_Maine,_2000

- http://www.270towin.com/states/Maine

- http://www.pressherald.com/2016/11/08/mainers-head-to-polls-in-historic-election/

- http://mainepublic.org/post/maine-voter-turnout-extremely-heavy#stream/0

- http://www.csmonitor.com/USA/Elections/2012/1106/Voter-turnout-the-6-states-that-rank-highest-and-why/Maine

- http://www.nonprofitvote.org/documents/2010/10/maine-voter-turnout-2008-pdf.pdf

- http://cdn.bipartisanpolicy.org/wp-content/uploads/sites/default/files/2012%20Voter%20Turnout%20Full%20Report.pdf

- https://www.census.gov/prod/2006pubs/p20-556.pdf

- https://www.census.gov/prod/2010pubs/p20-562.pdf

- https://www.census.gov/prod/2011pubs/12statab/election.pdf

- https://www.deptofnumbers.com/gdp/maine/

Now, upload the dataset. From SAS Studio, click Refresh, then import it, as we explained in Chapter 2. Change the name of the dataset from IMPORT to CORRELATION.

Click on Tasks and Utilities ➤ Tasks ➤ Statistics ➤ Correlation Analysis. Select in Data the dataset CORRELATION. In Analysis variables, select GDP. In Correlate with, choose voterturnout, as shown in Figure 4-12.

Figure 4-12. *Correlation analysis of GDP and voter turnout in Maine*

In Options ➤ Plots, select Individual Scatter Plots, as in Figure 4-13.

Figure 4-13. *The Options tab*

Then, click Run. The output will be as shown in Figure 4-14.

1 With Variables:	voterturnout
1 Variables:	gdp

Pearson Correlation Coefficients, N = 5	
	gdp
voterturnout voterturnout	0.04995

Figure 4-14. *Pearson correlation coefficients*

The correlation coefficient is 0.05, which indicates that there is almost no relation between the GDP and voter turnout in Maine.

On the plot, you will find the points are all over the place, as shown in Figure 4-15.

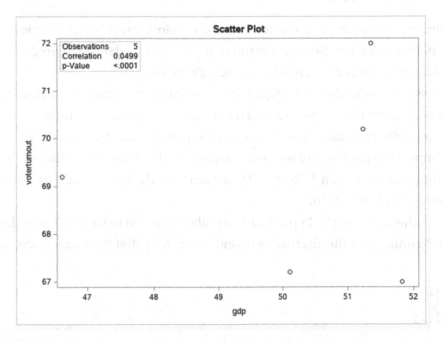

Figure 4-15. *Correlation between voter turnout and GDP*

Be aware that this is not the general rule. In your state, the correlation might be different. The state of Maine historically has one of the highest voter turnouts. I think that the weather might have a larger impact than the GDP. However, this is just a theory and needs further investigation. As an exercise, you could build a dataset of the temperature and voter turnout in Maine and verify if there is a high correlation between the weather and voter turnout. Further, you could investigate your state and see which factors affect voter turnout there.

T-Tests

Often, we are interested in knowing whether the means of two independent groups are truly different or if the observed difference merely occurred by chance.

The T-test is used to determine the probability of the difference between two sets of data. In simpler words, it compares the mean values between two samples.

Note that the T-test is used for small samples because they will not follow the normal distribution.

One-Sample T-tests

In this example, I collected all the votes from the previous presidential elections in Maine for both parties' candidates over the years. I used this dataset in predicting the results in Maine for both the 2016 and the 2020 elections.

The dataset name is MaineVotesDR.xlsx. It has four columns: Year of elections, Democrat voters, Republican voters, and the Result. The Result column is either 0 or 1. If it is 0, the Republican candidate won. If it is 1, the Democratic candidate won.

Now, import the dataset and let us see how to use the T-test to see the results in 2020. In Tasks, click Statistics, then T Tests. In Data, select the dataset. In the Analysis variable, select Result, as in Figure 4-16.

Our hypothesis is that the Republican candidate will win in the 2020 presidential election in Maine, while the alternative hypothesis will be that the Democratic candidate will win.

```
HO: Result=0
Ha: Result=1
```

Figure 4-16. *T-test*

In the code, the default value of the null hypothesis is zero. You will find it in the Code tab as h0=0. Do not change the default value, as it as our hypothesis. Now, click Run and check the p-value that SAS Studio generates in the reports, as in Figure 4-17.

Variable: Result (Result)

Tests for Normality				
Test		**Statistic**		**p Value**
Shapiro-Wilk	W	0.630341	Pr < W	<0.0001
Kolmogorov-Smirnov	D	0.384889	Pr > D	<0.0100
Cramer-von Mises	W-Sq	0.454844	Pr > W-Sq	<0.0050
Anderson-Darling	A-Sq	2.680494	Pr > A-Sq	<0.0050

Variable: Result (Result)

N	Mean	Std Dev	Std Err	Minimum	Maximum
15	0.6000	0.5071	0.1309	0	1.0000

Mean	95% CL Mean		Std Dev	95% CL Std Dev	
0.6000	0.3192	0.8808	0.5071	0.3713	0.7997

DF	t Value	Pr > |t|
14	4.58	0.0004

Figure 4-17. *T-test results: Prediction for 2020 presidential election in Maine*

The p-value = 0.0004. Since the p-value is less than 0.05, we can reject the null hypothesis. Hence, we can conclude that the Republican candidate will lose in Maine in the 2020 election.

Paired-sample T-test

Now, we will test whether the difference between price and cost is significantly different from 30. In other words, the null hypothesis will be that the difference between the price and cost equals 30, while the alternative hypothesis of the difference will not be equal to 30.

HO: Difference=30.
Ha: Difference ^= 30 or in other languages syntax is: Ha: Difference !=30.

To do this example, click on Tasks and Utilities ➤ Tasks ➤ Statistics ➤ T Tests. Select in Data the dataset PRICEDDATA in SASHELP. In T-test type, select Paired test. In Group 1 Variable, choose price, and for Group 2 Variable, choose cost, as in Figure 4-18.

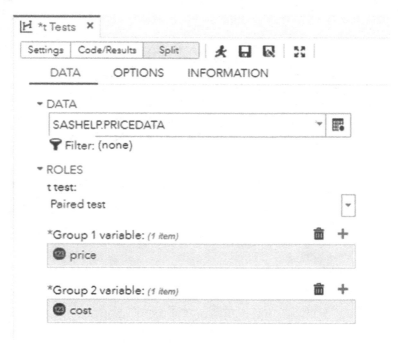

Figure 4-18. *Paired T-test Data tab*

Now, click the Options tab. In Tests ➤ Tails, select Two-tailed test. In the Alternative hypothesis, put mu1-mu2 ^=30, as in Figure 4-19. You have the choice whether to include the tests for normality and the default plots. Usually, it is better to include them and present them in your reports. However, in this example, we mainly focus on the value of the p-value in the results.

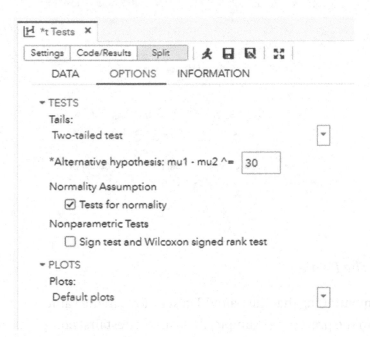

Figure 4-19. *The Options tab*

In the results, the p-value is less than 0.0001, as in Figure 4-20. Hence, we have enough evidence to reject the null hypothesis, because the p-value is less than 0.05.

Variable: _Difference_ (Difference: price - cost)

Tests for Normality				
Test		Statistic	p Value	
Shapiro-Wilk	W	0.896986	Pr < W	<0.0001
Kolmogorov-Smirnov	D	0.1888	Pr > D	<0.0100
Cramer-von Mises	W-Sq	7.159388	Pr > W-Sq	<0.0050
Anderson-Darling	A-Sq	39.28743	Pr > A-Sq	<0.0050

Difference: price - cost

N	Mean	Std Dev	Std Err	Minimum	Maximum
1020	42.0448	21.9813	0.6883	6.5700	93.4000

Mean	95% CL Mean		Std Dev	95% CL Std Dev	
42.0448	40.6942	43.3954	21.9813	21.0671	22.9791

DF	t Value	Pr > |t|
1019	17.50	<.0001

Figure 4-20. *The P-value*

It is worth mentioning that the paired T-test is done in the same way as the one-sample T-test. In the preceding example, the paired T-test first computes the difference between the price and the cost, then performs the one-sample T-test over the difference, as in the previous section.

Two-Sample T-tests

In this example, the question is: Is there a significant difference in height between the two genders?

The null hypothesis is that there is no difference in height. The alternate hypothesis is that there is a difference in height based on gender.

In this example, choose the CLASS dataset from SASHELP. Choose the Two-sample test in the T test. Finally, select Height as the Analysis variable and Sex as the Groups variable, as in Figure 4-21.

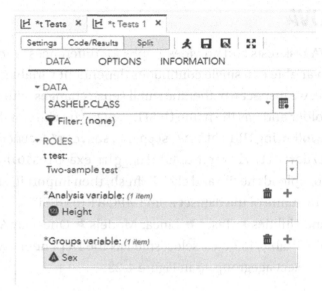

Figure 4-21. *T-test Data tab*

Click Run and check the p-value in the output, as in Figure 4-22.

Equality of Variances				
Method	Num DF	Den DF	F Value	Pr > F
Folded F	8	9	1.03	0.9527

Figure 4-22. *P-value*

We cannot reject the null hypothesis as there is not enough evidence for that; the p-value is larger than 0.05. We can conclude that there is no difference in height between the two genders. However, we must note that the sample is small. The SASHELP.CLASS dataset contains only nineteen rows.

Linear Models

ANOVA is short for ANalysis Of VAriance. A one-way ANOVA is used to test the difference between the means of two groups on a single variable. The T-test does this as well and is used more frequently than the ANOVA. However, some clients insist on asking for the ANOVA test with the T-test to compare both outputs for additional verification.

One-Way ANOVA

The one-way ANOVA task tests and provides graphs for differences among the means of a single categorical variable or a single continuous dependent variable.

For this example, we inspect whether the smell sense changes with age. The dataset is in the "Dataset" folder and has the name UPSIT.csv. Or, you can run the code to create the dataset from the following URL: `http://support.sas.com/documentation/cdl/en/statug/66859/HTML/default/viewer.htm#statug_glm_examples10.htm`.

From SAS Studio, upload the file and click Refresh, then import it, as we explained in Chapter 2. Change the name of the dataset from IMPORT to UPSIT.

Click on Tasks and Utilities ➤ Tasks ➤ Linear Models ➤ One-Way ANOVA. In Data, select the WORK.UPSIT table, in Roles, select smell as the Dependent variable, and in Categorical variable, select agegroup, as in Figure 4-23.

Figure 4-23. *One-Way ANOVA Data tab*

Click Run. The output will be as in Figures 4-24 to 4-28. The figures indicate that the sense of smell significantly differs across the age groups. Figure 4-24 shows that there are five age groups in this sample, with 180 observations. As usual, the null hypothesis is that there is no relationship between the olfactory sense and age. The alternate hypothesis is that there is a relationship between the two variables. The p-values, in Figure 4-24, are all less than 0.05, which indicates that there is a relationship between the sense of smell and age.

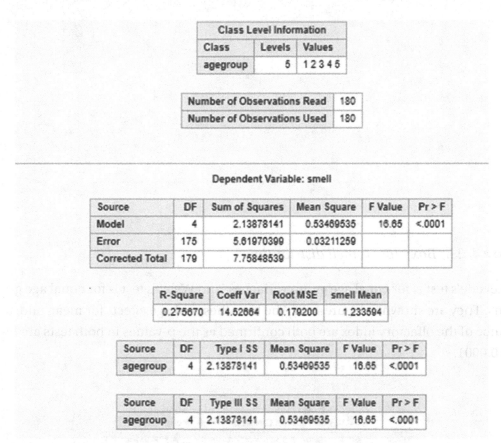

Figure 4-24. *Class-level information and p-values*

Figure 4-25 shows clearly different degrees of variability for the olfactory index within different age groups, with the variability generally rising with age. The sense of smell weakens and decreases with older age.

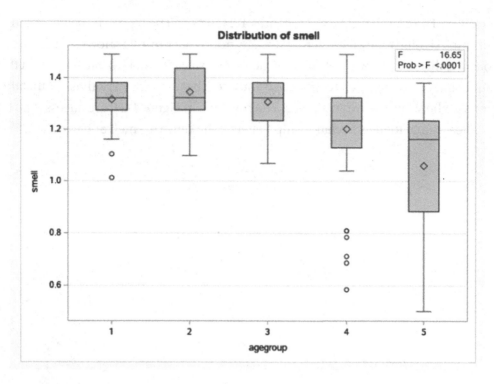

Figure 4-25. *Box plot of smell distribution*

Levene's test is for equal age group variances, and Welch's test is for equal age group means. They are shown in Figure 4-26. The hypotheses of age effects for mean and variance of the olfactory index are both confirmed as the p-values in both tests are less than 0.0001.

Levene's Test for Homogeneity of smell Variance ANOVA of Squared Deviations from Group Means					
Source	DF	Sum of Squares	Mean Square	F Value	Pr > F
agegroup	4	0.0799	0.0200	6.35	<.0001
Error	175	0.5503	0.00314		

Welch's ANOVA for smell			
Source	DF	F Value	Pr > F
agegroup	4.0000	13.72	<.0001
Error	78.7489		

Figure 4-26. *Levene's and Welch's tests*

Figure 4-27 shows no remarkable difference between the means of the first three groups, while the difference increases with groups four and five.

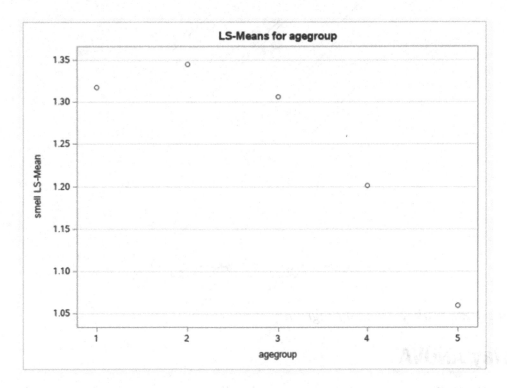

Figure 4-27. *LS-Means for age group*

The Tukey diagram in Figure 4-28 shows the mean difference between each of the groups. There is no significant difference between the first three groups, which are shown in the figure with red lines. However, there is a significant difference between the first three groups and groups four and five, which are shown in the figure with blue lines. The Tukey output is interactive in SAS Studio; roll your mouse over every line, and SAS Studio shows which two groups it is comparing.

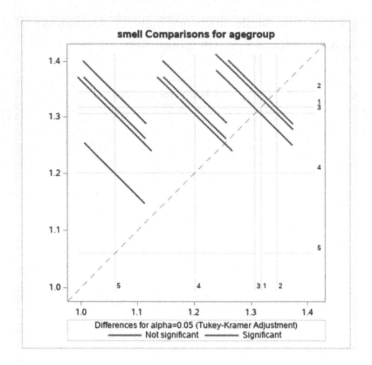

Figure 4-28. *Tukey-Kramer diagram*

N-Way ANOVA

The N-Way ANOVA task provides graphs for the effects of one or more factors on the means of a single, continuous dependent variable.

In this example, assume that a flight company wants to run an advertising campaign. It needs to know which customers to target to increase its revenue. Therefore, we investigate if there is a significant difference in revenue based on the type and the source of the revenue.

Select Tasks and Utilities ➤ Tasks ➤ Linear Models ➤ N-Way ANOVA. In the data, select the SASHELP.REVHUB2 table. In Roles, select Revenue as the Dependent variable and Source and Type as the Factors, as in Figure 4-29.

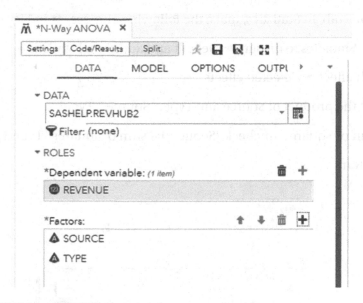

Figure 4-29. *N-Way ANOVA Data tab*

Click on the Model tab. Click on Edit beside the Intercept. Add Source and Type. To add their product, select them using CTRL + A or by selecting them using the mouse. Then, click on Full Factorial, as in Figure 4-30.

Model Effects Builder

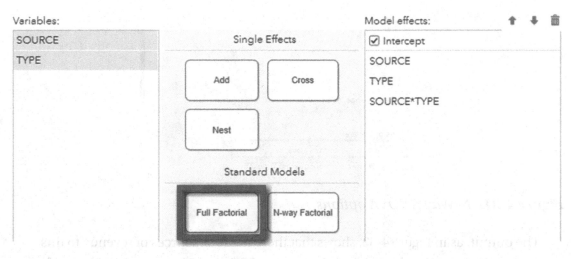

Figure 4-30. *Model effects builder*

Scroll down and click OK.

In the Options tab, in Statistics, make the following choices, as in Figure 4-31:

- Select Statistics to display ➤ Default and additional statistic

- Which effect ➤ Selected effects

- Select the product of Source and Type: "Source * Type"

- In Sum of Squares, uncheck "Sequential sum of squares (Type 1)"

Then, click Run.

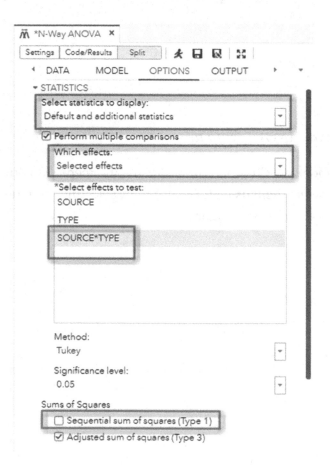

Figure 4-31. *N-Way ANOVA options*

The output, as in Figure 4-32, shows that there are four sources of revenue to this company, which are Freight, Other, Passenger, and Service. There are three types of revenue, which are Direct, Indirect, and Other.

Class Level Information		
Class	Levels	Values
SOURCE	4	Freight Other Passenger Service
TYPE	3	Direct Indirect Other

Number of Observations Read	72
Number of Observations Used	72

Figure 4-32. *Class-level information*

The interaction plot for revenue, as shown in Figure 4-33, shows that the highest revenue comes from Passenger-Direct, followed by Freight-Direct, then Passenger-Indirect, then the rest of source*type. Roll the mouse over every point; SAS Studio will display the source, type, and revenue value.

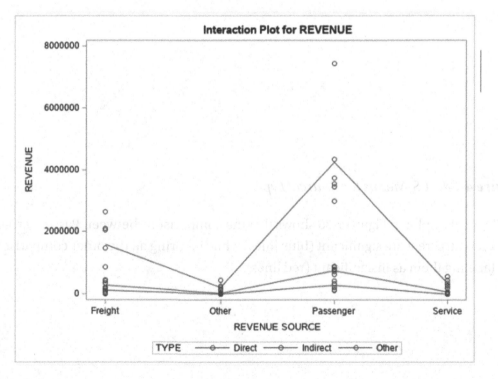

Figure 4-33. *Interaction plot for revenue*

Again, in Figure 4-34, the LS-Means for Source*Type also shows that Passenger-Direct is way higher than the other sources. Its value is more than $4 million, followed by Freight-Direct, with a value of about $1.6 million, and Passenger-Indirect, which is less than $1 million.

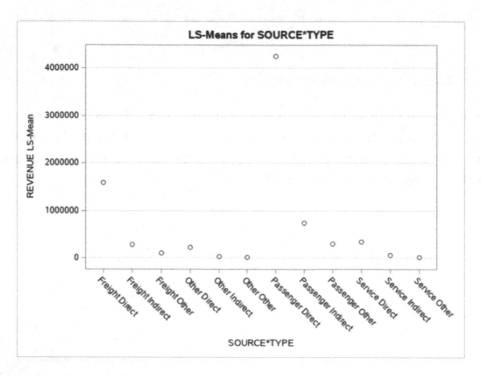

Figure 4-34. *LS-Means for Source*Type*

The Tukey plot in Figure 4-35 shows that the comparisons between Passenger-Direct and Freight-Direct are significant (blue lines) while ignoring all the other comparisons and labeling them as insignificant (red lines).

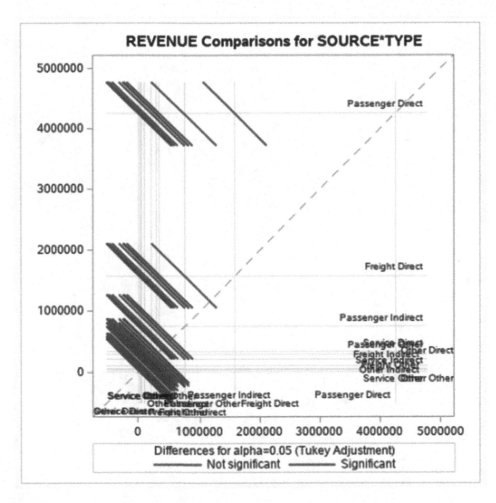

Figure 4-35. *Tukey-Kramer diagram for revenue comparison for Source*Type*

Summary

This chapter explains the most crucial five statistics concepts that will be needed for almost all analytics and data science reports. The five concepts are one-way frequency, summary statistics, correlation analysis, T-tests, and ANOVA. Moreover, the chapter introduces plenty of advanced visualizations and tables, such as Leven and Welch tables and Tukey and LS-Means plots.

Again, we did all these tasks while relying primarily on the IDE and explaining the code as well. Now, it is time to go more intensely into code. In the next chapter, we shall step deeper into SAS programming and learn the operators, the loops, and the SELECT statement, and touch on PROC SQL.

CHAPTER 5

Advanced Data Preprocessing and Feature Engineering

We mentioned in the preface that in this book we try to balance between using the integrated development environment (IDE) and the code. In Chapters 5 and 6, we focus on the code. This chapter shows some advanced data cleaning and querying methods, such as using PROC SQL, SELECT, WHERE, WHEN, and DO loops.

Comment Statement

It is crucial to document your code so you and your peers can understand the reasoning behind how the analysis is constructed. In general, if you do not document your code, you might look at it after a few months and not remember why you used a certain function or why you made a certain check. Hence, code documentation is crucial so as to follow the logic flow.

The syntax of the comment statement in SAS Studio can be done in two ways as follows:

```
*message;
/*message*/
```

You can use the comment statement anywhere in a SAS program to explain your code. The SAS compiler ignores text in comment statements during processing.

It is always recommended that you start your programs by writing a comment at the top explaining what your program will do, and then add as many comments as you need. To describe the logic flow, it is best to write a comment at every data or procedure step to describe what it does.

© Engy Fouda 2020

E. Fouda, *Learn Data Science Using SAS Studio*, https://doi.org/10.1007/978-1-4842-6237-5_5

Arithmetic Operators

Arithmetic operators indicate that an arithmetic calculation is performed, as shown in Table 5-1.

Table 5-1. *Arithmetic Operators[1]*

Symbol	Definition	Example	Result
**	exponentiation	a**3	raise A to the third power
*	Multiplication*	2*y	multiply 2 by the value of Y
/	division	var/5	divide the value of VAR by 5
+	addition	num+3	add 3 to the value of NUM
-	subtraction	sale-discount	subtract the value of DISCOUNT from the value of SALE

* The asterisk (*) is always necessary to indicate multiplication; 2Y and 2(Y) are not valid expressions.

If a missing value is an operand for an arithmetic operator, the result is a missing value. You have to handle the missing values in your dataset by using an IF statement to avoid propagating missing values. We discuss the IF statement in detail in Chapter 6.

How to Represent Missing Values in Raw Data

Table 5-2 shows how to represent each type of missing value in raw data, so that SAS reads and stores the value appropriately.

---

[1]https://documentation.sas.com/?docsetId=lrcon&docsetTarget=p00iah2thp63bmn1lt20esa
g14lh.htm&docsetVersion=9.4&locale=en

Table 5-2. *Missing Values Conventions*[2]

Missing Values	Representation in Data
Numeric	. (a single decimal point)
Character	' ' (a blank enclosed in quotation marks)
Special	.letter (a decimal point followed by a letter; for example, .B)
Special	._ (a decimal point followed by an underscore)

You can set values to missing within your data step by using program statements such as this one:

```
if age<0 then
age=.;
```

This statement sets the stored value of Age to a numeric missing value, if Age has a value less than 0.

For more information about handling missing values, please check the SAS documentation at the following link:

```
http://bit.ly/2tX68BO
```

The order of operations in SAS is as follows: parentheses, exponents, multiplication, division, addition, subtraction.

Comparison Operators

Comparison operators set up a comparison, operation, or calculation with two variables, constants, or expressions. If the comparison is true, the result is 1. If the comparison is false, the result is 0.

Comparison operators can be expressed as symbols or with their mnemonic equivalents, which are shown in Table 5-3. SAS Studio can process the symbol or the mnemonic. The mnemonic can be uppercase or lowercase or mixed case.

---

[2]https://documentation.sas.com/?docsetId=lrcon&docsetTarget=p175x77t7k6kggn1io94yed
qagl3.htm&docsetVersion=9.4&locale=en

Table 5-3. *Comparison Operators[3]*

Symbol	Mnemonic Equivalent	Definition	Example
=	EQ	equal to	a=3
^=	NE	not equal to[1]	a ne 3
¬=	NE	not equal to	
~=	NE	not equal to	
>	GT	greater than	num>5
<	LT	less than	num<8
>=	GE	greater than or equal to[2]	sales>=300
<=	LE	less than or equal to[3]	sales<=100
	IN	equal to one of a list	num in (3, 4, 5)

Again, you can use either the mnemonic or the symbol. For example:

```
if grade >= 50 then status = pass;
else status = fail;
```

similar to:

```
if grade GE 50 then status EQ pass;
else status EQ fail;
```

To learn more about SAS operators in expressions, please check the SAS documentation at this URL: http://bit.ly/36UGEnk.

PROC SQL Statement

Writing SQL statements is crucial for database developers. SAS Studio enables database developers to use PROC SQL statements where they can create tables, query them, select or delete or update any data from them, and merge or subset the tables as

---

[3]https://documentation.sas.com/?docsetId=lrcon&docsetTarget=p00iah2thp63bmn1lt20esa
g14lh.htm&docsetVersion=9.4&locale=en#n10lfrm906gpv7n1t7fue0g1ayqz

developers used to do. Throughout this book, we show plenty of examples of various ways to achieve those tasks using other SAS statements. In this section, you will learn how to use PROC SQL.

The sequence for the PROC SQL statement is as follows:

```
PROC SQL;
    select
    from
    group by
    having
    order by
;
quit;
```

To remember this sequence, you can use this acronym: "So Few Workers Go Home On Time." The source of this acronym is https://blogs.sas.com/content/sastraining/2012/04/24/go-home-on-time-with-these-5-proc-sql-tips/.

In Listing 2-5 in Chapter 2, we created a new table in the newlib library by filtering out the cars that were made in Europe. In Listing 5-1, we do the same task but by using PROC SQL.

Listing 5-1. PROC SQL Example

```
proc sql;
create table newlib.europeancars as
select * from sashelp.cars where origin = "Europe" ;
quit;
```

In this example, Listing 5-1, we start with PROC SQL. Then, we create a table and use the AS keyword to define the table contents. We use SELECT * to select all the columns from SASHELP.CARS. For the condition to filter the European cars only, we use the WHERE clause and check on the value of the origin. If the origin = Europe, the row is added to the table and excluded otherwise.

SELECT-WHERE Statement

The SELECT statement is the primary tool of PROC SQL. You use it to identify, retrieve, and manipulate columns of data from a table.

The following simple SELECT statement is sufficient to produce a useful result:

```
select party
   from maine.voters;
```

The preceding statement contains a SELECT clause that lists the party column and a FROM clause that lists the table in which the party column resides.

WHERE Clause

The WHERE clause adds more specification to SELECT statement lists by adding a condition to it. It is similar to the IF-THEN-ELSE statement that we shall talk about in the next chapter.

For example, if we want to check the party distribution for seniors in Maine, we should add a WHERE clause to the SELECT statement as follows:

```
select Party
   from maine.voters;
   where age >= 65;
```

Another example, Listing 5-2, is from the sample questions of the SAS Base Programming Certificate Free Preparation Guide.

Listing 5-2. WHERE Clause Example

```
data class;
input name $ gender $ age;
datalines;
Anna F 23
Ben M 25
Bob M 21
Brian M 27
Edward M 26
Emma F 32
```

```
Joe M 34
Sam F 32
Tom M 24
;
data WORK.MALES_OVER25;
set WORK.CLASS;
where Gender="M";
where age>25;
run;
```

How many observations are in the dataset WORK.MALES_OVER25?

The output will be five observations. The output is Figure 5-1. The table contains both males and females whose age is larger than 25, because only the last WHERE clause took effect.

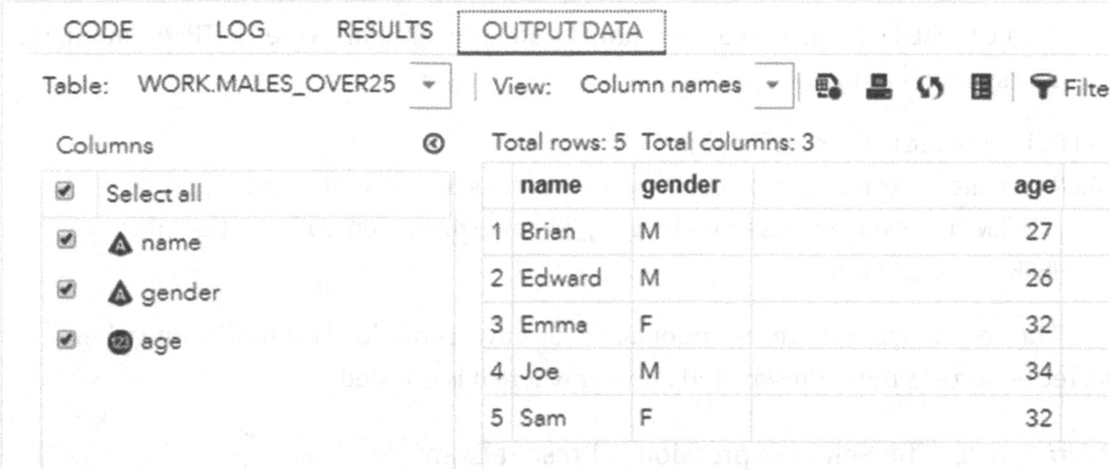

CODE	LOG	RESULTS	OUTPUT DATA

Table: WORK.MALES_OVER25 ▼ | View: Column names ▼   🖳 📇 ↻ 📊 | 🍷 Filte

Columns	◎	Total rows: 5 Total columns: 3			
☑ Select all		**name**	**gender**		**age**
☑ ⚠ name		1 Brian	M		27
☑ ⚠ gender		2 Edward	M		26
☑ ㉓ age		3 Emma	F		32
		4 Joe	M		34
		5 Sam	F		32

Figure 5-1. *The output of WHERE clause example*

If you want to modify the code to have a table that contains males whose age is larger than 25, you should combine both conditions with the AND operator.

Instead of two WHERE statements:

```
where Gender="M";
where age>25;
```

It should be:

```
where Gender="M" and age>25;
Or use WHERE and WHERE ALSO statements, as follows:
```

```
where Gender="M";
where also age>25;
```

In Listing 5-2, the WHERE statement is used to subset the dataset. There is another way to subset the dataset, which is by using the IF statement without THEN; we will have an example about it in the next chapter.

Moreover, this listing shows that if we have several consecutive WHERE statements, not WHERE ALSO, only the last one will take effect.

SELECT-WHEN-OTHERWISE Statement

The SELECT-WHEN statement is a conditional statement similar to the IF-THEN statement.

The syntax is as follows:

```
SELECT <(select-expression)>;
WHEN-1 (when-expression-1 <..., when-expression-n>) statement;
<... WHEN-n (when-expression-1 <..., when-expression-n>) statement;>
<OTHERWISE statement;>
```

The select-expression is optional. Let us have a couple of examples, one where select-expression is present and another where it is omitted.

Listing 5-3. The Select-Expression Is Present Example

```
DATA MYDATA;
SET sashelp.class;
SELECT(age);
 when (10) status="child";
 WHEN (11,12) status="preteen";
 OTHERWISE status="teenager";
END;
PROC PRINT DATA=MYDATA;
RUN;
```

If `select-expression` is present, SAS evaluates the `select-expression` and `when-expression`. SAS compares the two for equality and returns a value of true or false. If the comparison is true, the statement is executed. If the comparison is false, execution proceeds to the next `when-expression`. If no `WHEN` statements remain, the `OTHERWISE` statement will be executed. See Figure 5-2.

Obs	Name	Sex	Age	Height	Weight	status
1	Alfred	M	14	69.0	112.5	teenager
2	Alice	F	13	56.5	84.0	teenager
3	Barbara	F	13	65.3	98.0	teenager
4	Carol	F	14	62.8	102.5	teenager
5	Henry	M	14	63.5	102.5	teenager
6	James	M	12	57.3	83.0	preteen
7	Jane	F	12	59.8	84.5	preteen
8	Janet	F	15	62.5	112.5	teenager
9	Jeffrey	M	13	62.5	84.0	teenager
10	John	M	12	59.0	99.5	preteen
11	Joyce	F	11	51.3	50.5	preteen
12	Judy	F	14	64.3	90.0	teenager
13	Louise	F	12	56.3	77.0	preteen
14	Mary	F	15	66.5	112.0	teenager
15	Philip	M	16	72.0	150.0	teenager
16	Robert	M	12	64.8	128.0	preteen
17	Ronald	M	15	67.0	133.0	teenager
18	Thomas	M	11	57.5	85.0	preteen
19	William	M	15	66.5	112.0	teenager

Figure 5-2. *The Output of Select-Expression Clause Example*

If there is no `select-expression`, the `when-expression` is evaluated to produce a Boolean result. If the result is true, the statement is executed. If the result is false, SAS proceeds to the next `when-expression` or to the `OTHERWISE` statement. If the result of all `when-expressions` is false and no `OTHERWISE` statement is present, SAS issues an error message. If more than one `WHEN` statement has a true `when-expression`, only the first `WHEN` statement is used. See Figure 5-3.

Listing 5-4. No Select-Expression Example

```
DATA MYDATA;
SET sashelp.class;

SELECT;
 WHEN (age=15 and sex="F") research="include";
 OTHERWISE research="not include";
END;
PROC PRINT DATA=MYDATA;
RUN;
```

Obs	Name	Sex	Age	Height	Weight	research
1	Alfred	M	14	69.0	112.5	not include
2	Alice	F	13	56.5	84.0	not include
3	Barbara	F	13	65.3	98.0	not include
4	Carol	F	14	62.8	102.5	not include
5	Henry	M	14	63.5	102.5	not include
6	James	M	12	57.3	83.0	not include
7	Jane	F	12	59.8	84.5	not include
8	Janet	F	15	62.5	112.5	include
9	Jeffrey	M	13	62.5	84.0	not include
10	John	M	12	59.0	99.5	not include
11	Joyce	F	11	51.3	50.5	not include
12	Judy	F	14	64.3	90.0	not include
13	Louise	F	12	56.3	77.0	not include
14	Mary	F	15	66.5	112.0	include
15	Philip	M	16	72.0	150.0	not include
16	Robert	M	12	64.8	128.0	not include
17	Ronald	M	15	67.0	133.0	not include
18	Thomas	M	11	57.5	85.0	not include
19	William	M	15	66.5	112.0	not include

Figure 5-3. *The Output of the Absent Select-Expression Clause Example*

DO Loops

There are three types of DO loops for iteration: DO LOOP, DO UNTIL, and DO WHILE. The basic syntax for the DO loop is as follows:

```
DO indexvariable= initialvalue to endvalue     <by incrementvalue>;
    SAS statements . . .
END;
```

By default, the increment value is 1, but you can change it. The following example is to add the odd numbers only starting from 1 to 10. To be able to save the iteration in the output dataset, use the OUTPUT statement; this saves every observation instantly instead of at the end of the data step. See Figure 5-4.

Listing 5-5. DO LOOP Example

```
data b;
SUM=0;
DO ILOOP=1 to 10 by 2;
  SUM=SUM + ILOOP;
  output;* The OUTPUT statement tells SAS to write the current observation
  to a SAS data set immediately, not at the end of the DATA step;
END;

run;
```

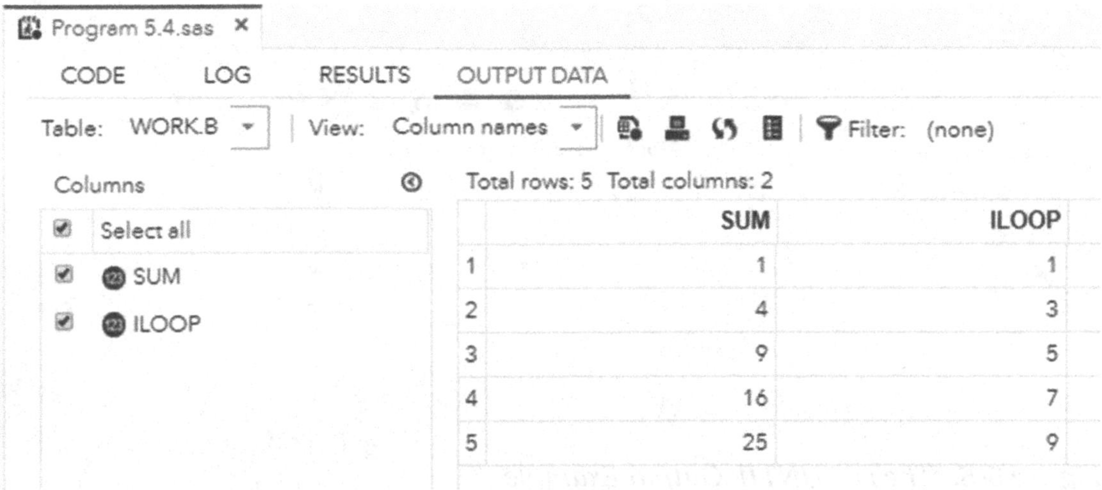

Figure 5-4. *The DO LOOP Output Example*

The DO UNTIL loop keeps iterating until the condition of the UNTIL clause becomes true. The UNTIL condition is evaluated at the end of the loop. Let's redo the previous example using the DO UNTIL loop. Pay attention that you must increment the counter manually inside the loop; otherwise, it is an infinite loop. The good news is that even if you did an infinite loop, there is no need to worry, as SAS Studio won't crash. It will take an longer time than normal in execution, then it will generate an error for you. Hence, again, it is always a good habit to check the SAS log even if you receive results. See Listing 5-6 and Figure 5-5.

Listing 5-6. DO UNTIL Example

```
data d;
SUM=0;
i=1;
DO UNTIL (i GT 9);
  SUM=SUM + i;
  output;
  i=i+2;

END;

run;
```

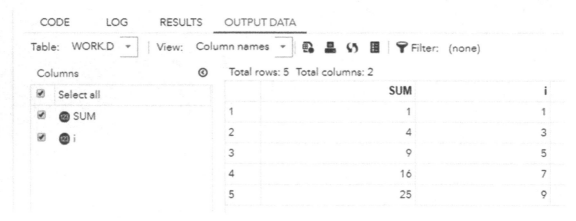

Figure 5-5. *The DO UNTIL Output Example*

The `DO While` is the inverse of the `DO UNTIL`. The loop continues until the condition in the `WHILE` clause is false. See Listing 5-7 and Figure 5-6.

Listing 5-7. DO WHILE Example

```
data e;
SUM=0;
i=1;
```

```
DO WHILE(i LT 10);
  SUM=SUM + i;
  output;
  i=i+2;

END;

run;
```

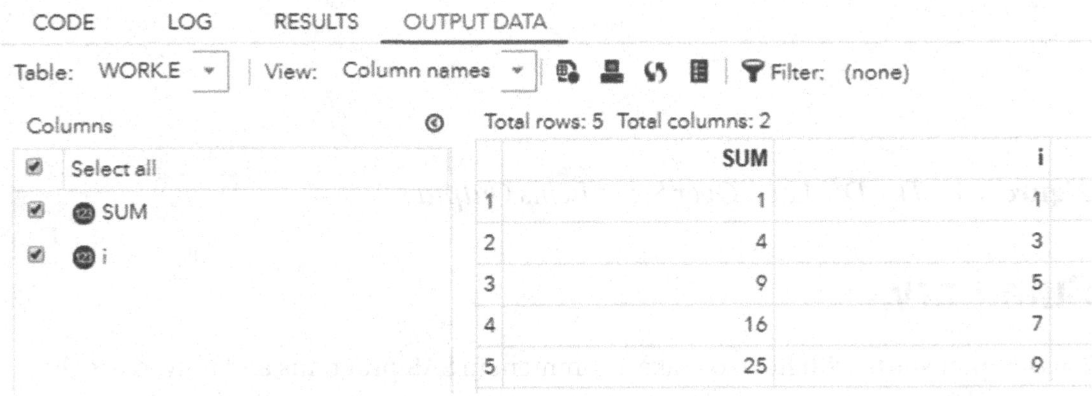

	SUM	i
1	1	1
2	4	3
3	9	5
4	16	7
5	25	9

Figure 5-6. *The DO WHILE Output Example*

One last way to do the DO loops is by looping over a set of items if there is no specific pattern for the iteration counter. We still can get the same result, as in the previous examples, but by using a set of numbers instead of using a counter and increment statement. See Listing 5-8 and Figure 5-7.

Listing 5-8. DO Loop Over Set of Items Example

```
data f;
SUM=0;
DO i=1,3,5,7,9;
  SUM=SUM + i;
  output;
END;

run;
```

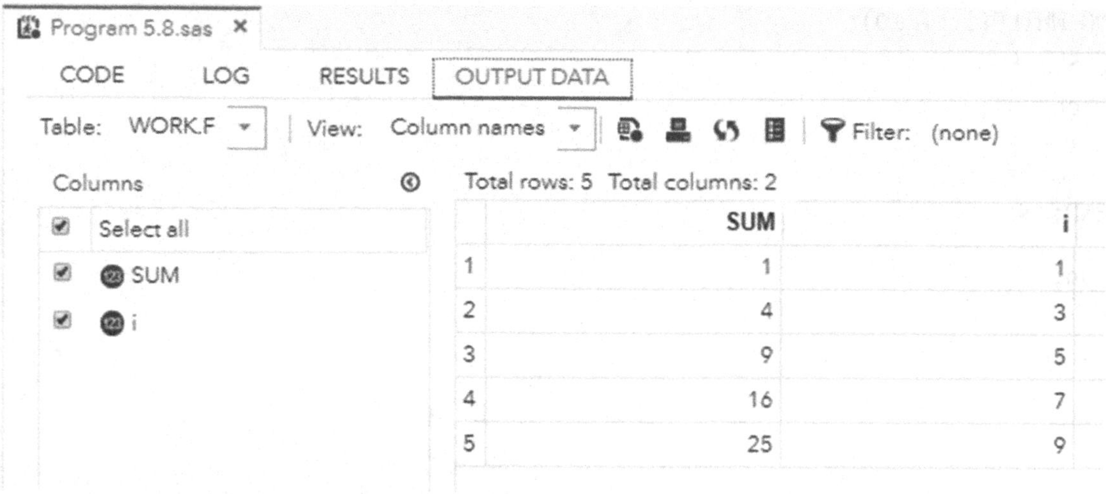

Figure 5-7. *The DO Loop Over Set of Items Output*

Summary

This chapter starts with how to make a comment in SAS programs and how to use the arithmetic and comparison operators. We touched on PROC SQL and discussed the SELECT statement in detail. The SELECT statement has various combinations with WHERE and WHEN statements. Further, we discussed the loops in SAS in detail. There are DO loops, and DO UNTIL and DO WHILE loops. In the next chapter, we shall learn how to prepare for analysis and learn the essential conditional statement.

CHAPTER 6

Preparing Data for Analysis

In this chapter, we continue digging into SAS programming, still focusing on the code rather than the GUI. As we mentioned earlier, in any programming language, you learn the data types, operators, IF condition, and loops. We covered all these topics in Chapters 2 and 5, except the IF condition statements. In this chapter, we learn about them and also more advanced data preparation and processing techniques.

Label

In Chapter 2, we mentioned that the SAS program consists of two parts: the data step and the proc step. In the previous chapters, we tried many procedures; for example, PROC SGPLOT and PROC MEANS. The LABEL statement lies in the data step, and the LABEL option lies in PROC PRINT.

The first step to analyze the dataset is to understand the variables that you have. Identify which are the dependent variables and which are the independent variables, which are numeric/continuous, and which are character. Usually, your client will provide you with this information in a text file along with the file that contains the data. The text file is called the dictionary because it contains the explanations of the variable names and their types. Let us recall from Chapter 1, Figure 1-2, the files that the Secretary of State of Maine office sent me.

The data is in the Excel sheet, and the dictionary is highlighted in the red box in the middle. The variable names of YOB, ENROLL, and DT Accept are ambiguous without the dictionary. From the dictionary, we understand what each of these variables means; for example, YOB is Year Of Birth.

To append these explanatory names to the dataset in SAS, we use the LABEL statement. The dataset is in the "Dataset" folder and has the name Voter_A.CSV. From SAS Studio, upload it and click Refresh, then import it, as we explained in Chapter 2. Change the name of the dataset from IMPORT to Voters.

© Engy Fouda 2020
E. Fouda, *Learn Data Science Using SAS Studio*, https://doi.org/10.1007/978-1-4842-6237-5_6

This dataset is not a real dataset. I created it using the same variables that were sent to me by the Secretary of Maine. However, this file does not contain real information.

After importing the file, we find that it has one hundred rows and nine columns, as in Figure 6-1.

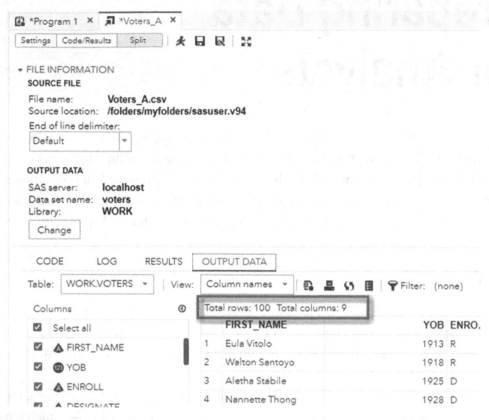

Figure 6-1. *Import Voters_A.CSV*

Now, let us write a LABEL statement in the data step and use the LABEL option of PROC PRINT, as in Listing 6-1.

Listing 6-1. Label Statement and Label Option

```
data voters_label;
set work.voters;
label
FIRST_NAME='Name'
YOB='Year Of Birth'
```

```
ENROLL='Enrollment Code'
DESIGNATE='Special Designations'
DT_ACCEPT='Date Accepted (Date of Registration)'
CG='Congressional District'
CTY='County ID'
DT_CHG='Date Changed'
DT_LAST_VPH='Date Of Last Statewide Election with VPH';
run;

proc print label;
run;
```

When you run Listing 6-1, the output is as in Figure 6-2.

Obs	Name	Year Of Birth	Enrollment Code	Special Designations	Date Accepted (Date of Registeration)	Congressional District	County ID	Date Changed	Date Of Last Statewide Election wwith VPH
1	Eula Vitolo	1913	R		13123	2	01AND	11/26/2008	
2	Walton Santoyo	1918	R		17418	2	01AND	05/22/2008	06/10/2008
3	Aletha Stabile	1925	D		19281	2	01AND	05/17/2010	11/02/2010
4	Nannette Thong	1928	D		38664	2	01AND	04/25/2012	11/04/2008
5	Courtney Bonner	1929	R		40106	2	01AND	06/13/2012	06/14/2016
6	Leda Glessner	1932	R		25132	2	01AND	12/31/2005	11/06/2012
7	Cammie Sidoti	1935	R		38337	2	01AND	08/18/2010	11/04/2014
8	Regina Bundy	1944	R		26150	2	01AND	02/08/2016	11/04/2014
9	Trisha Tuch	1949	D		38845	2	01AND	11/07/2007	06/14/2016
10	Vonda Delahoussaye	1950	D		32449	2	01AND	03/29/2016	11/06/2012
11	Fransisca Schroder	1963	U		39251	2	01AND	12/30/2015	11/04/2008
12	Flavia Adolphson	1996	D		41078	2	01AND	07/25/2012	.
13	Jarrod Sells	1995	D		42346	2	01AND	12/18/2015	.
14	Wallace Wohl	1949	U		41193	2	01AND	10/24/2012	
15	Elijah Ledezma	1980	U		42221	2	01AND	08/12/2015	
16	Vertie Stack	1964	U		34500	2	01AND	12/31/2005	11/04/2014
17	Verena Sapienza	1938	R		23648	2	01AND	12/31/2005	06/14/2016
18	Jarvis Stancill	1950	U		39006	2	01AND	12/31/2005	11/04/2014
19	Carina Benson	1955	D		36017	2	01AND	09/08/2016	11/04/2014
20	Hillary Naber	1960	U		01/01/1850	2	01AND	12/07/2009	11/04/2008
21	Daisey Scherer	1966	U		36567	2	01AND	05/15/2012	11/04/2014
22	Kurtis Howle	1947	R		38292	2	01AND	12/31/2008	11/04/2014

Figure 6-2. *Output with labels, not variable names*

Figure 6-2 shows that the report is printed with descriptive labels and not the variable names. Hence, it is more readable. However, the data is clearly unclean. "Special Designations" is a blank column, and there are missing values in other columns. Probably, the data needs verification that it is valid and does not contain any wrong information. In the rest of this chapter's sections, you will learn how to clean such messy data.

Format

In the previous section, we saw how the LABEL statement changed the output to a more readable format. On the other hand, the FORMAT statement changes the input, not the output. We mentioned in Chapter 2, in the "Variable Types" section, that, by default, the length of the character data type is eight characters. Hence, if we use the INPUT statement instead of importing the Excel file, the FORMAT statement overrides the default variables' formats. Moreover, the FORMAT statement can assign the same format to several variables at one time.

The dataset is in the "Dataset" folder and has the name Voters_A.CSV. From SAS Studio, upload it and click Refresh. Do not import it, because we shall use the INPUT statement, as we explained the last section of Chapter 2.

Run Listing 6-2 and check the output. In Listing 6-2, we still did not use the FORMAT statement.

Listing 6-2. Read CSV File with the INPUT Statement

```
data voters_format;
infile "/folders/myfolders/SAS4Days/Voters_A.csv" dlm=',' firstobs=2;
input NAME $ YOB $  ENROLL $ DESIGNATE $ DT_ACCEPT $  CG  CTY $ DT_CHG
$  DT_LAST_VPH $;
run;

proc print ;
run;
```

In Listing 6-2, the first line, DATA VOTERS_FORMAT creates a new dataset in the Work library with the name VOTERS_FORMAT. In the second line, the INFILE statement reads the CSV file from the assigned path, sets the delimiter to comma, and tells SAS Studio that the first observation is at the second row in the CSV file, because the file has a header. The third line uses the INPUT statement to set the variables' names and their types. To distinguish between the character and numeric variables, you put a dollar sign ($) after the character variables, so SAS Studio can read them correctly. The last line in the data step is the RUN statement to end it. Finally, PROC PRINT prints the dataset in a printable report.

Run the program and check the "Name" column. SAS Studio truncates the names and reads only the first eight characters. For example, the first name in VOTERS_A.CSV is "Eula Vitolo," while SAS Studio truncates it to "Eula Vit," as in Figure 6-3. As we said in Chapter 2, SAS Studio, by default, assigns eight characters only to the character field.

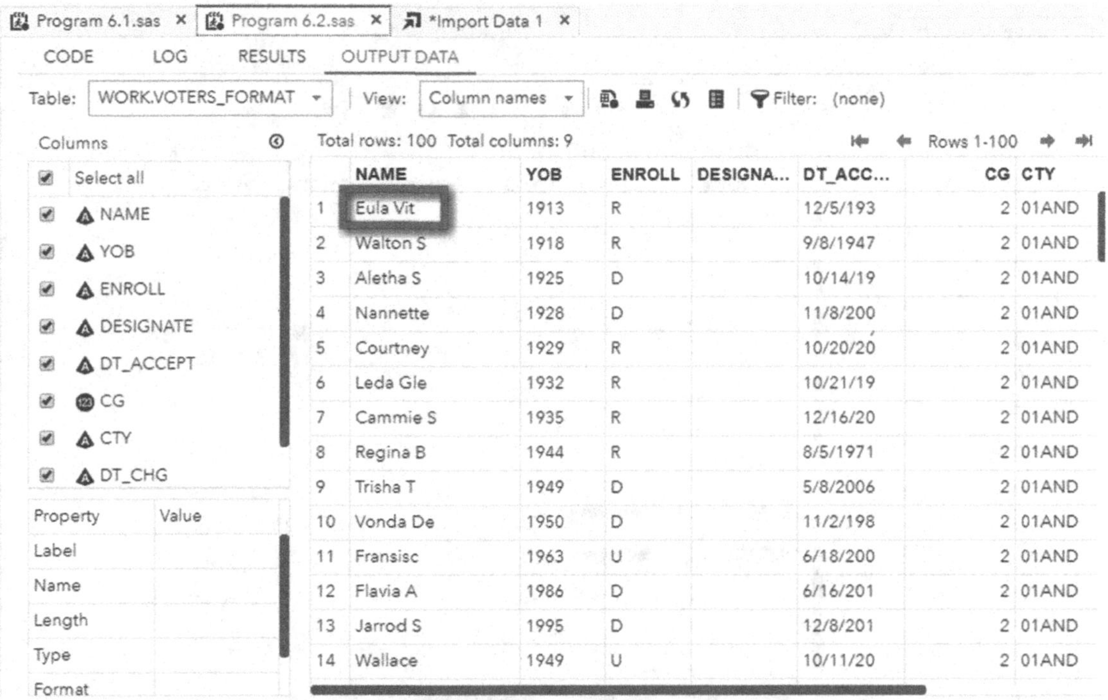

Figure 6-3. *Name values are truncated to eight characters only*

To override this rule and resolve this problem, we use the FORMAT statement to read all the characters, as in Listing 6-3.

Listing 6-3. FORMAT Statement

```
data voters_format;
Format Name $21. ;
infile "/folders/myfolders/SAS4Days/Voters_A.csv" dlm=',' firstobs=2;
input NAME $ YOB $  ENROLL $ DESIGNATE $ DT_ACCEPT $  CG  CTY $ DT_CHG
$  DT_LAST_VPH $;

run;
proc print ;
run;
```

After running Listing 6-3, the names are displayed entirely, as in Figure 6-4. The FORMAT statement specifies the type of the NAME variable as character by using the $ sign, and the number of characters is 21. Remember to put the dot at the end ($21.), so SAS Studio can read the values correctly.

Figure 6-4. *The names are complete now*

Create New Variables

In SAS, you do not need to define a new variable beforehand or set its data type as you do in some other languages. You just put the name of the new variable and its value. That is it. SAS will be able to identify the data type of the new variable by itself.

In this example, let us continue using the dataset of the previous sections, Voters_A. CSV. In the dataset, there is a variable called YOB, which is the Year Of Birth. The new variable that we will create is the age of the voters.

As in Listing 6-4, the age equals the difference between the current year and the birth date. The age is defined as age=2020-YOB. We define it in the data step.

Listing 6-4. Creating New Variables

```
data voters_format;
Format Name $21. ;
infile "/folders/myfolders/SAS4Days/Voters_A.csv" dlm=',' firstobs=2;
input NAME $ YOB $  ENROLL $ DESIGNATE $ DT_ACCEPT $  CG  CTY $ DT_CHG
$  DT_LAST_VPH $;

*creating age as a new variable;
age=2020-YOB;
run;

proc print ;
run;
```

After running Listing 6-4, the output has a new variable appended to it, as in Figure 6-5.

Obs	Name	YOB	ENROLL	DESIGNATE	DT_ACCEPT	CG	CTY	DT_CHG	DT_LAST_VPH	age
1	Eula Vitolo	1913	R		12/5/193	2	01AND	11/26/20		106
2	Walton Santoyo	1918	R		9/8/1947	2	01AND	5/22/200	6/10/200	101
3	Aletha Stabile	1925	D		10/14/19	2	01AND	5/17/201	11/2/201	94
4	Nannette Thong	1928	D		11/8/200	2	01AND	4/25/201	11/4/200	91
5	Courtney Bonner	1929	R		10/20/20	2	01AND	6/13/201	6/14/201	90
6	Leda Glessner	1932	R		10/21/19	2	01AND	12/31/20	11/6/201	87
7	Cammie Sidoti	1935	R		12/16/20	2	01AND	8/18/201	11/4/201	84
8	Regina Bundy	1944	R		8/5/1971	2	01AND	2/8/2016	11/4/201	75
9	Trisha Tuch	1949	D		5/8/2006	2	01AND	11/7/200	6/14/201	70
10	Vonda Delahoussaye	1950	D		11/2/198	2	01AND	3/29/201	11/6/201	69
11	Fransisca Schroder	1963	U		6/18/200	2	01AND	12/30/20	11/4/200	56
12	Flavia Adolphson	1986	D		6/16/201	2	01AND	7/25/201		33
13	Jarrod Sells	1995	D		12/8/201	2	01AND	12/18/20		24
14	Wallace Wohl	1949	U		10/11/20	2	01AND	10/24/20		70
15	Elijah Ledezma	1980	U		8/5/2015	2	01AND	8/12/201		39
16	Vertie Stack	1964	U		6/15/199	2	01AND	12/31/20	11/4/201	55
17	Verena Sapienza	1938	R		9/28/196	2	01AND	12/31/20	6/14/201	81
18	Jarvis Stancill	1950	U		10/16/20	2	01AND	12/31/20	11/4/201	69
19	Carina Benson	1955	D		8/10/199	2	01AND	9/8/2016	11/4/201	64
20	Hillary Naber	1960	U		01/01/18	2	01AND	12/7/200	11/4/200	59

Figure 6-5. *Output of Listing 6-4*

Scroll down a little and check the age. There is one voter listed as being 340 years old, seen in Figure 6-6. Probably, this value is wrong. We see later in this chapter how to resolve this issue by using the IF statement.

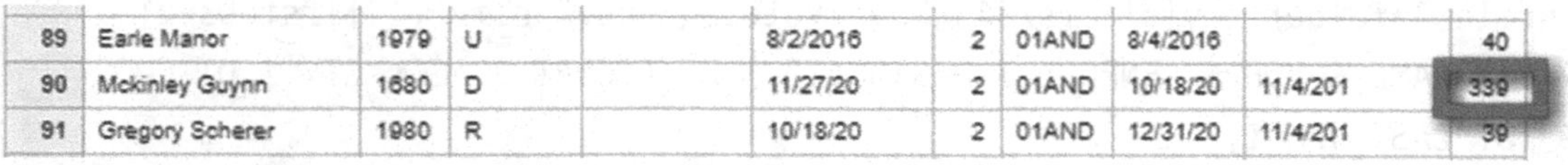

89	Earle Manor	1979	U		8/2/2016	2	01AND	8/4/2016		40
90	Mckinley Guynn	1680	D		11/27/20	2	01AND	10/18/20	11/4/201	339
91	Gregory Scherer	1980	R		10/18/20	2	01AND	12/31/20	11/4/201	39

Figure 6-6. *Check the age values for wrong ones*

Rearrange the Dataset Variables

Before cleaning the data, you might like to reorder the variables in the dataset; for example, rearrange the variables to have the age variable directly beside YOB. We use RETAIN and SET statements to achieve that.

However, it has to be done in a new data step, as in Listing 6-5. Later in this chapter, we shall learn more about the SET statement.

Listing 6-5. Retain Statement

```
data voters_format;
Format Name $21. ;
infile "/folders/myfolders/SAS4Days/Voters_A.csv" dlm=',' firstobs=2;
input NAME $ YOB $  ENROLL $ DESIGNATE $ DT_ACCEPT $  CG  CTY $ DT_CHG
$  DT_LAST_VPH $;

*creating age as a new variable;
age=2020-YOB;
run;

data voters_retain;
retain NAME YOB age ENROLL DESIGNATE DT_ACCEPT  CG  CTY DT_CHG  DT_LAST_VPH;
set voters_format;
run;
proc print data=voters_retain;
run;
```

After running Listing 6-5, age now is after YOB, as in Figure 6-7. If you got an error, please check the file path in the INFILE statement. Change it to where the file is located on your machine.

Obs	NAME	YOB	age	ENROLL	DESIGNATE	DT_ACCEPT	CG	CTY	DT_CHG	DT_LAST_VPH
1	Eula Vitolo	1913	107	R		12/5/193	2	01AND	11/26/20	
2	Walton Santoyo	1918	102	R		9/8/1947	2	01AND	5/22/200	6/10/200
3	Aletha Stabile	1925	95	D		10/14/19	2	01AND	5/17/201	11/2/201
4	Nannette Thong	1928	92	D		11/8/200	2	01AND	4/25/201	11/4/200
5	Courtney Bonner	1929	91	R		10/20/20	2	01AND	6/13/201	6/14/201
6	Leda Glessner	1932	88	R		10/21/19	2	01AND	12/31/20	11/6/201
7	Cammie Sidoti	1935	85	R		12/16/20	2	01AND	8/18/201	11/4/201
8	Regina Bundy	1944	76	R		8/5/1971	2	01AND	2/8/2016	11/4/201
9	Trisha Tuch	1949	71	D		5/8/2006	2	01AND	11/7/200	6/14/201
10	Vonda Delahoussaye	1950	70	D		11/2/198	2	01AND	3/29/201	11/6/201
11	Fransisca Schroder	1963	57	U		6/18/200	2	01AND	12/30/20	11/4/200
12	Flavia Adolphson	1986	34	D		6/16/201	2	01AND	7/25/201	
13	Jarrod Sells	1995	25	D		12/8/201	2	01AND	12/18/20	
14	Wallace Wohl	1949	71	U		10/11/20	2	01AND	10/24/20	
15	Elijah Ledezma	1980	40	U		8/5/2015	2	01AND	8/12/201	
16	Vertie Stack	1964	56	U		6/15/199	2	01AND	12/31/20	11/4/201
17	Verena Sapienza	1938	82	R		9/28/196	2	01AND	12/31/20	6/14/201
18	Jarvis Stancill	1950	70	U		10/16/20	2	01AND	12/31/20	11/4/201
19	Carina Benson	1955	65	D		8/10/199	2	01AND	9/8/2016	11/4/201
20	Hillary Naber	1960	60	U		01/01/18	2	01AND	12/7/200	11/4/200
21	Daisey Scherer	1966	54	U		2/11/200	2	01AND	5/15/201	11/4/201
22	Kurtis Howle	1967	53	R		11/1/200	2	01AND	12/31/20	11/4/201

Figure 6-7. *Voters dataset after reordering its variables to have age after YOB*

There are other ways to reorder the variables in the dataset. You can check them at this link:

http://support.sas.com/kb/8/395.html

IF Statement

Now, this is the crucial section for data cleaning—the IF statement. There are three forms in which to use the IF statement:

1. IF (Condition) without THEN statement

2. IF-THEN statement

3. IF-THEN-ELSE statement

IF (Condition) Without THEN statement

The IF statement without a THEN statement is used to subset the datasets. In this example, let us continue using the dataset of the previous sections, Voters_A.CSV. We want to have two new datasets as subsets of the original, where the first one contains only voters who are enrolled as Democrat, and the other contains voters who are enrolled as Republican. We use an IF statement and check the ENROLL variable value, as in Listing 6-6.

Use the SET statement to copy the old dataset to a new one. It is usually better to create a new copy of the dataset and do all your analysis over the new dataset and keep the original one intact. This is a good practice, especially for those who use datasets provided by governments or the United Nations or banks.

Listing 6-6. IF (Condition) Statement Example

```
data voters_enroll;
Format Name $21. ;
infile "/folders/myfolders/sasuser.v94/Voters_A.csv" dlm=',' firstobs=2;
input NAME $ YOB $  ENROLL $ DESIGNATE $ DT_ACCEPT $  CG  CTY $ DT_CHG
$  DT_LAST_VPH $;
run;

data voters_democrat;
set voters_enroll;
if enroll='D';
run;

data voters_republican;
set voters_enroll;
if enroll='R';
run;

proc print data=voters_enroll (obs=3);
run;

proc print data=voters_democrat (obs=3);
run;
proc print data=voters_republican (obs=3);
run;
```

After running Listing 6-6, we have three datasets. Voters_Enroll has all the parties, Voters_Democrats has the Democratic voters only, and Voters_Republicans has the Republican voters only. We need three data steps, one for each dataset; similarly, we need three PROC PRINTs. If you got an error, please check the file path in the INFILE statement. Change it to where the file is located on your machine.

Instead of printing all the rows in the dataset, we added (OBS=3) option to the PROC PRINT to print three rows only from each dataset. Click on the Results tab to see the three reports, as in Figure 6-8.

Obs	Name	YOB	ENROLL	DESIGNATE	DT_ACCEPT	CG	CTY	DT_CHG	DT_LAST_VPH
1	Eula Vitolo	1913	R		12/5/193	2	01AND	11/26/20	
2	Walton Santoyo	1918	R		9/8/1947	2	01AND	5/22/200	6/10/200
3	Aletha Stabile	1925	D		10/14/19	2	01AND	5/17/201	11/2/201

Obs	Name	YOB	ENROLL	DESIGNATE	DT_ACCEPT	CG	CTY	DT_CHG	DT_LAST_VPH
1	Aletha Stabile	1925	D		10/14/19	2	01AND	5/17/201	11/2/201
2	Nannette Thong	1928	D		11/8/200	2	01AND	4/25/201	11/4/200
3	Trisha Tuch	1949	D		5/8/2006	2	01AND	11/7/200	6/14/201

Obs	Name	YOB	ENROLL	DESIGNATE	DT_ACCEPT	CG	CTY	DT_CHG	DT_LAST_VPH
1	Eula Vitolo	1913	R		12/5/193	2	01AND	11/26/20	
2	Walton Santoyo	1918	R		9/8/1947	2	01AND	5/22/200	6/10/200
3	Courtney Bonner	1929	R		10/20/20	2	01AND	6/13/201	6/14/201

Figure 6-8. *Print three rows after subsetting the dataset*

Click on the Output Data tab and select the table from the drop-down menu, as in Figure 6-9. WORK.VOTERS_ENROLL contains all the parties and has 100 rows.

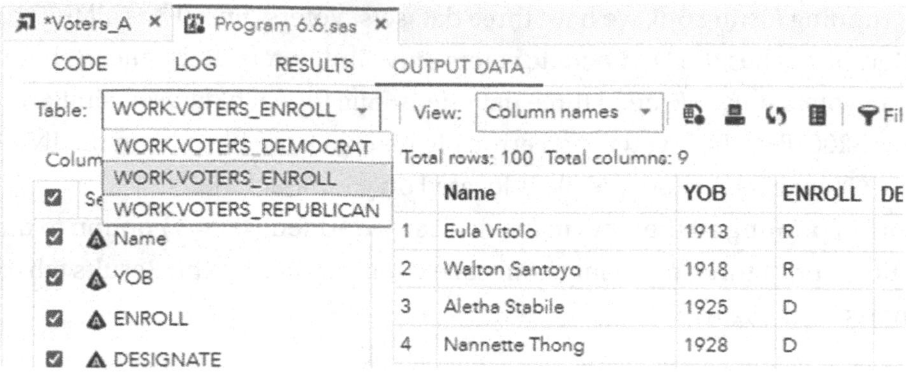

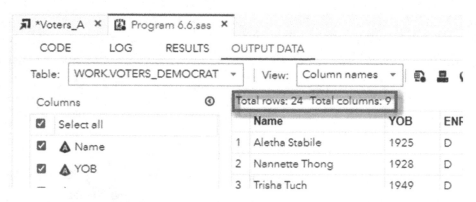

Figure 6-9. *Select the output table from the drop-down menu*

Select the WORK.VOTERS_DEMOCRAT table, which has 24 rows only, as in Figure 6-10.

Figure 6-10. *Democrat table*

Select the WORK.VOTERS_REPUBLICAN table, which has 24 rows only, as in Figure 6-11.

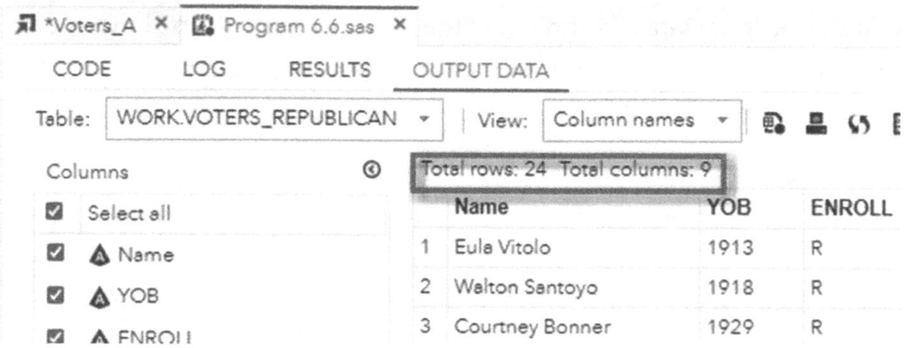

Figure 6-11. *Republican table*

To know the rest of the parties that we have in the table and how many rows and the percentage of each one, run one-way frequencies, as in Figure 6-12. Return to the dictionary of the party codes:

A   AMERICANS ELECT

D   DEMOCRATIC

G   GREEN INDEPENDENT

L   LIBERTARIAN

O   OTHER QUALIFYING PARTY

R   REPUBLICAN

U   UNENROLLED

X   INELIGIBLE

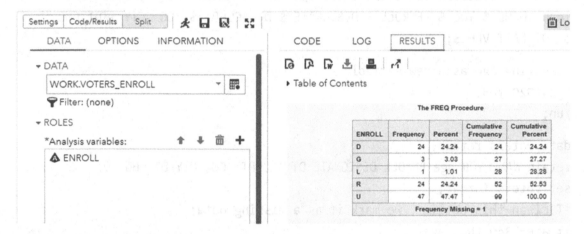

Figure 6-12. *One-way frequencies output*

IF-THEN Statement

As mentioned earlier, the "Age" column has a definite wrong value for one of the voters, where they are 340 years old. To correct this, if the dataset data was a real one, we would have had several options:

1.   Get the correct age by contacting the Secretary of Maine office or contacting the voter.

2.   Remove the value and put a dot instead of it to indicate that it is a missing data, as mentioned in Chapter 5. The SAS convention for missing data is to put a dot for the missing numeric value and a blank for the missing character value.

3. Another solution is to delete the whole row.

4. Leave the value as it is. However, this value will be an outlier and will affect the analysis when trying to find the relationship between the age, party, and the expected voting.

In this example, consider it as a missing value. To do so, we use the IF-THEN statement, as in Listing 6-7.

Listing 6-7. IF-THEN Statement Example

```
data voters_format;
Format Name $21. ;
infile "/folders/myfolders/sasuser.v94/Voters_A.csv" dlm=',' firstobs=2;
input NAME $ YOB $  ENROLL $ DESIGNATE $ DT_ACCEPT $  CG  CTY $ DT_CHG
$  DT_LAST_VPH $;

*creating age as a new variable;
age=2020-YOB;
run;

data voters_retain;
retain NAME YOB age ENROLL DESIGNATE DT_ACCEPT  CG  CTY DT_CHG  DT_LAST_VPH;
set voters_format;
*T clean this outlier, we mark it as a missing data;
if age= 340 then age=.;
run;

proc print data=voters_retain;
run;
```

After running Listing 6-7, the value will be missing, as in Figure 6-13.

89	Earle Manor	1979	40	U		8/2/2016	2	01AND	8/4/2016	
90	Mckinley Guynn	1680	.	D		11/27/20	2	01AND	10/18/20	11/4/201
91	Gregory Scherer	1980	39	R		10/18/20	2	01AND	12/31/20	11/4/201

Figure 6-13. *Make the wrong value a missing value*

The other solution is to delete the whole row, as in Listing 6-8, using the DELETE statement.

Listing 6-8. Delete the Whole Row of the Wrong Age Value

```
data voters_format;
Format Name $21. ;
infile "/folders/myfolders/sasuser.v94/Voters_A.csv" dlm=',' firstobs=2;
input NAME $ YOB $  ENROLL $ DESIGNATE $ DT_ACCEPT $  CG  CTY $ DT_CHG
$  DT_LAST_VPH $;

*creating age as a new variable;
age=2020-YOB;
run;

data voters_retain;
retain NAME YOB age ENROLL DESIGNATE DT_ACCEPT  CG  CTY DT_CHG  DT_LAST_VPH;
set voters_format;
*Delete the whole row;
if age= 340 then Delete;
run;

proc print data=voters_retain;
run;
```

After running Listing 6-8, the whole row is deleted, as in Figure 6-14.

89	Earle Manor	1979	40	U		8/2/2016	2	01AND	8/4/2016	
90	Gregory Scherer	1980	39	R		10/18/20	2	01AND	12/31/20	11/4/201
91	Rob Heth	1980	39	R		11/4/201	2	01AND	11/6/201	11/4/201

Figure 6-14. *The row was deleted*

For further verification, compare the number of rows in the new dataset, WORK.
VOTERS_RETAIN, with those in the original dataset, WORK.VOTERS_FORMAT. In
Figure 6-15, the number of rows is 99, while in the original dataset, the number of rows is
100, as in Figure 6-16.

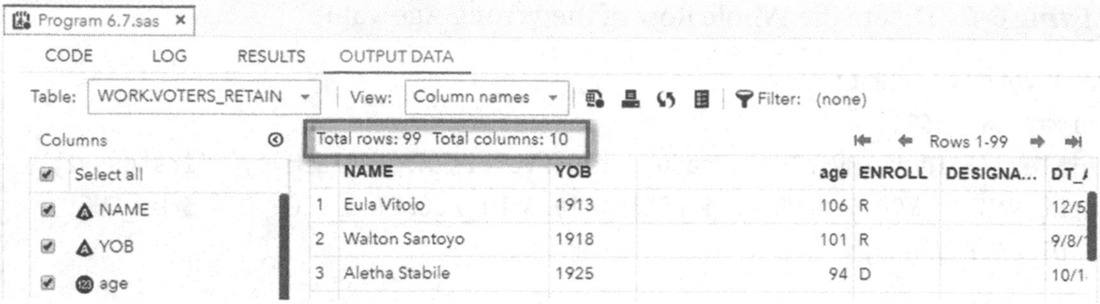

Figure 6-15. *New dataset after deleting the row*

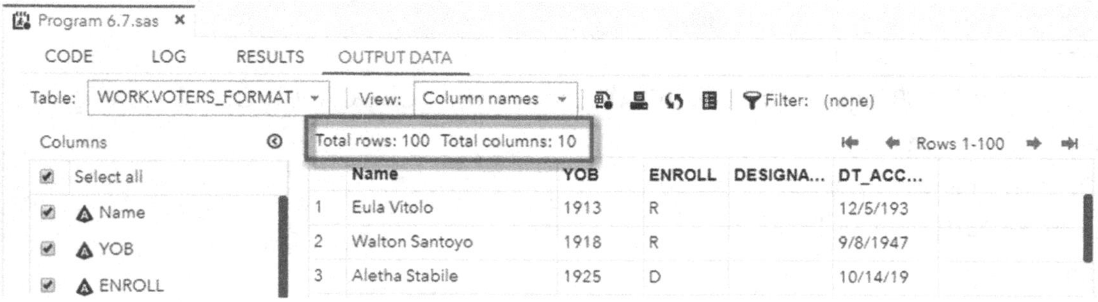

Figure 6-16. *The original dataset with 100 rows*

IF-THEN-ELSE Statement

Now, let us do an analysis by counting how many voters in the following age groups: 18–24, 25–29, 30–39, 40–49, 50–64, and 65+, as in Listing 6-9.

Listing 6-9. IF-THEN-ELSE Statement Example

```
data voters_format;
Format Name $21. ;
infile "/folders/myfolders/SAS4Days/Voters_A.csv" dlm=',' firstobs=2;
input NAME $ YOB $  ENROLL $ DESIGNATE $ DT_ACCEPT $  CG  CTY $ DT_CHG
$  DT_LAST_VPH $;

*creating age as a new variable;
age=2020-YOB;
run;
```

```
data voters_retain;
retain NAME YOB age ENROLL DESIGNATE DT_ACCEPT  CG  CTY DT_CHG  DT_LAST_VPH;
set voters_format;

if age= 340 then Delete;

if (age>=18 and age<=24) then age_group='18-24';
else if (age>=25 and age<=29) then age_group='25-29';
else if (age>=30 and age<=39) then age_group='30-39';
else if (age>=40 and age<=49) then age_group='40-49';
else if (age>=50 and age<=64) then age_group='50-64';
else if (age>=65) then age_group='65+';

run;

proc print data=voters_retain;

run;
```

After running Listing 6-9, the output is as in Figure 6-17.

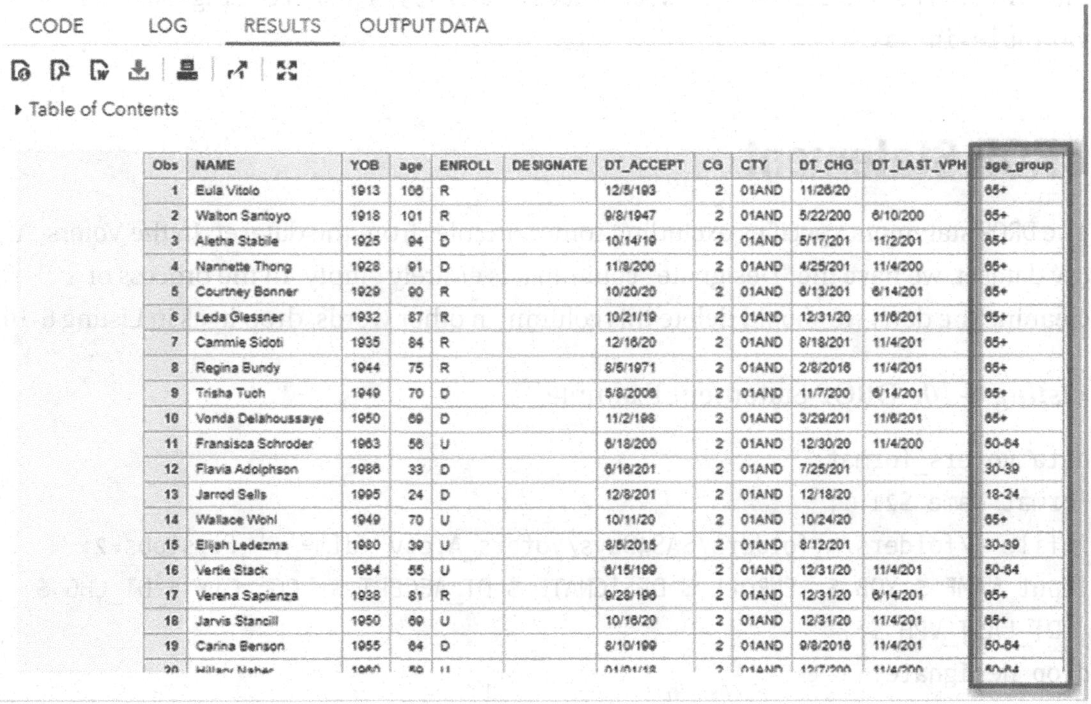

Obs	NAME	YOB	age	ENROLL	DESIGNATE	DT_ACCEPT	CG	CTY	DT_CHG	DT_LAST_VPH	age_group
1	Eula Vitolo	1913	106	R		12/5/193	2	01AND	11/26/20		65+
2	Walton Santoyo	1918	101	R		9/8/1947	2	01AND	5/22/200	6/10/200	65+
3	Aletha Stabile	1925	94	D		10/14/19	2	01AND	5/17/201	11/2/201	65+
4	Nannette Thong	1928	91	D		11/9/200	2	01AND	4/25/201	11/4/200	65+
5	Courtney Bonner	1929	90	R		10/20/20	2	01AND	6/13/201	6/14/201	65+
6	Leda Glessner	1932	87	R		10/21/19	2	01AND	12/31/20	11/6/201	65+
7	Cammie Sidoti	1935	84	R		12/16/20	2	01AND	8/18/201	11/4/201	65+
8	Regina Bundy	1944	75	R		8/5/1971	2	01AND	2/8/2016	11/4/201	65+
9	Trisha Tuch	1949	70	D		5/8/2006	2	01AND	11/7/200	6/14/201	65+
10	Vonda Delahoussaye	1950	69	D		11/2/198	2	01AND	3/29/201	11/6/201	65+
11	Fransisca Schroder	1963	56	U		6/18/200	2	01AND	12/30/20	11/4/200	50-64
12	Flavia Adolphson	1986	33	D		6/16/201	2	01AND	7/25/201		30-39
13	Jarrod Sells	1995	24	D		12/8/201	2	01AND	12/18/20		18-24
14	Wallace Wohl	1949	70	U		10/11/20	2	01AND	10/24/20		65+
15	Elijah Ledezma	1980	39	U		8/5/2015	2	01AND	8/12/201		30-39
16	Vertie Stack	1964	55	U		6/15/199	2	01AND	12/31/20	11/4/201	50-64
17	Verena Sapienza	1938	81	R		9/28/196	2	01AND	12/31/20	6/14/201	65+
18	Jarvis Stancill	1950	69	U		10/16/20	2	01AND	12/31/20	11/4/201	65+
19	Carina Benson	1955	64	D		8/10/199	2	01AND	9/8/2016	11/4/201	50-64
20	Hillary Naher	1960	59	U		01/01/18	2	01AND	12/7/200	11/4/200	50-64

Figure 6-17. Add AGE_GROUP variable

To get the count of each age group and its percentage, run one-way frequencies over the AGE_GROUP variable, as in Figure 6-18.

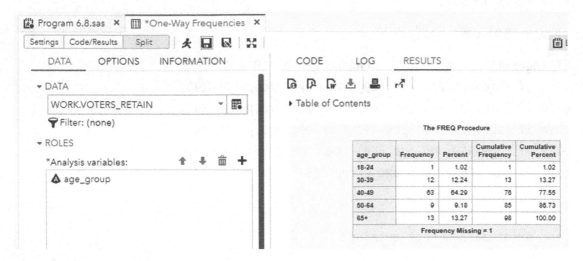

Figure 6-18.   *One-way frequencies over AGE_GROUP variable*

There are other ways to make AGE_GROUP in SAS. You can read more about them at this URL: `https://stackoverflow.com/questions/53395384/creating-an-agegroup-variable-in-sas`.

DROP Statement

The DROP statement specifies excluding some variables from the dataset. In the Voters_A. csv dataset, we have the "Designate" column. It is entirely empty. In the process of cleaning the data, we should delete this column; in other words, drop it, as in Listing 6-10.

Listing 6-10.   DROP Statement Example

```
data voters_format;
Format Name $21. ;
infile "/folders/myfolders/SAS4Days/Voters_A.csv" dlm=',' firstobs=2;
input NAME $ YOB $  ENROLL $ DESIGNATE $ DT_ACCEPT $  CG  CTY $ DT_CHG $
  DT_LAST_VPH $;
drop designate;
```

```
run;
proc print ;
run;
```

After running Listing 6-10, the output will have eight columns only, as in Figure 6-19.

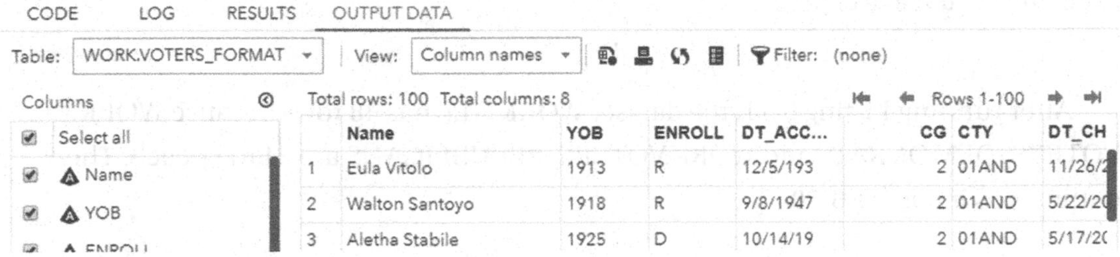

Figure 6-19. *Eight columns after dropping one*

SET Statement

We have been using the SET statement in the previous examples without discussing it in detail. We learned so far that we use the SET statement to copy an old dataset to a new one. Hence, all our processing is executed over the new dataset, while the old one is left intact. The SET statement has multiple functions. This website has many examples for each of these functions:

https://documentation.sas.com/?docsetId=lestmtsref&docsetTarget=p00hxg3x 8lwivcn1f0e9axziw57y.htm&docsetVersion=3.1&locale=en

However, in this section, we focus on appending two datasets into a new larger one. Simply, its syntax is: SET <dataset1> <dataset2> <dataset3> ....

Note that all the dataset names are separated by spaces.

We already have seen in Listing 6-6 how to copy the old dataset to a new dataset. In this example, let us do the reverse process. Assume we had the two Democrat and Republican datasets and wanted to combine them into one.

Re-run Listing 6-6 first, so you have the two datasets loaded in the Work library. Then, run Listing 6-11, where it merges the two datasets.

Listing 6-11. Appending Two Datasets

```
data all;
set voters_democrats voters_republicans;
run;

proc print data=all;
run;
```

After running Listing 6-11, the dataset WORK.ALL has 48 rows because WORK.
VOTERS_DEMOCRATS and WORK.VOTERS_REPUBLICANS are 24 rows each. The
output is as in Figure 6-20.

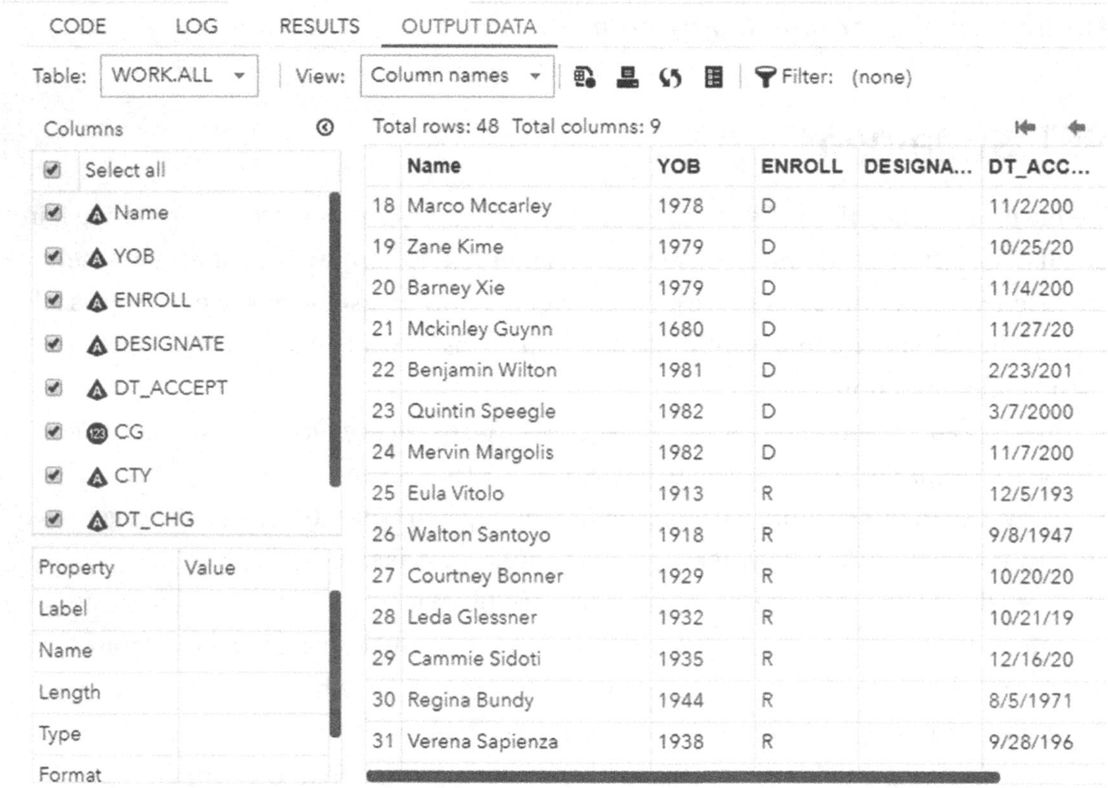

Figure 6-20. *Appending the Democrats and Republicans datasets*

Summary

This chapter explains the steps to clean and prepare data for analysis. It discusses the LABEL and FORMAT statements. It also explains how to create new variables and how to rearrange the variables in datasets. It explains the IF, IF-THEN, and IF-THEN-ELSE statements. Finally, it explains the DROP and SET statements. Now, we are done with the most common data analysis tasks, and the next chapter will introduce predicting the future using regression.

PART III

Advanced Topics

CHAPTER 7

Regression

The linear regression task fits a linear model to predict a single continuous dependent variable from one or more continuous or categorical predictor variables. This task produces statistics and graphs for interpreting the results.

Simple Linear Regression

In simple linear regression, there is only one independent variable in the model. The line equation will be as follows:

$$Y = a + b X$$

Where Y is the dependent variable that we want to predict its value, a is the intercept, b is the coefficient, and X is the independent variable. The error is ignored.

The null hypothesis is that there is no linear relationship between the variables. In other words, the value of b is zero. The alternative hypothesis is that there is a linear relationship, and b is not equal to zero.

H0: b = 0
Ha: b ^= 0

Let us see if there is a linear relationship between the prices of oil per barrel and gold.

The dataset is in the "Dataset" folder and has the name sp_oil_gold.xlsx. From SAS Studio, upload the file and click Refresh, then import it, as we explained in Chapter 2. Change the name of the dataset from IMPORT to sp_oil_gold.

The sources of this data are as follows:

- https://www.thebalance.com/presidential-elections-and-stock-market-returns-2388526

- https://www.statista.com/statistics/262858/change-in-opec-crude-oil-prices-since-1960/

© Engy Fouda 2020
E. Fouda, *Learn Data Science Using SAS Studio*, https://doi.org/10.1007/978-1-4842-6237-5_7

- https://www.statista.com/statistics/268027/change-in-gold-price-since-1990/

- https://www.nma.org/pdf/gold/his_gold_prices.pdf

- https://nma.org/wp-content/uploads/2018/02/his_gold_prices_1833_pres_2018.pdf

Click on Tasks and Utilities ➤ Tasks ➤ Linear Models ➤ Linear Regression. In Data, select WORK.SP_OIL_GOLD. In Dependent variable, select oil_price. In Continuous variables, select gold_prices, as in Figure 7-1.

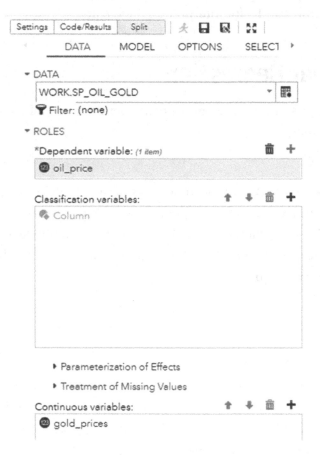

Figure 7-1. *Single linear regression*

Click on the Model tab and click Edit, as in Figure 7-2.

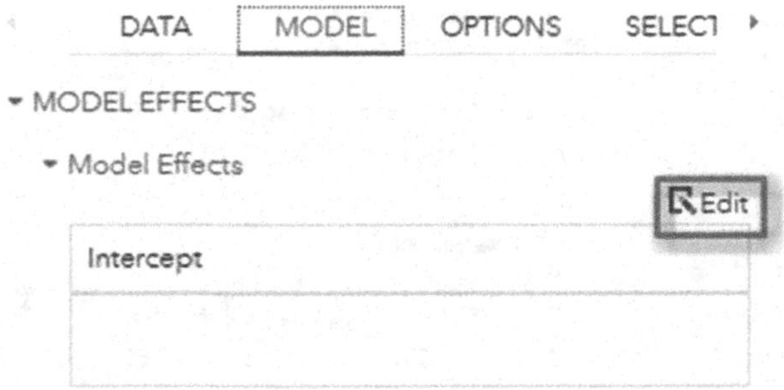

DATA   MODEL   OPTIONS   SELECT ▸

▾ MODEL EFFECTS

  ▾ Model Effects

                                      ⌐ Edit

    Intercept

Figure 7-2. *Model tab*

Select gold_prices and click Add, and then scroll down and click OK, as in Figure 7-3.

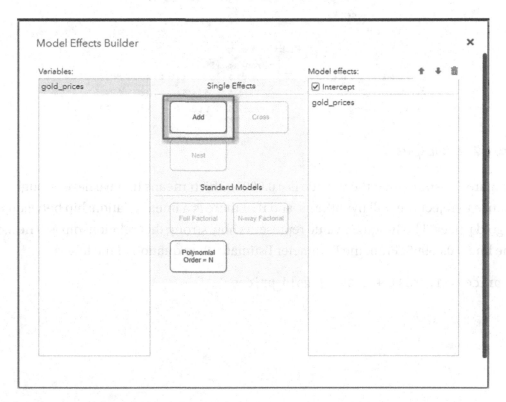

Figure 7-3. *Model effects builder*

Click Run. The output is shown in Figure 7-4.

Model: MODEL1
Dependent Variable: oil_price oil price

Number of Observations Read	15
Number of Observations Used	15

Analysis of Variance					
Source	DF	Sum of Squares	Mean Square	F Value	Pr > F
Model	1	11231	11231	42.56	<.0001
Error	13	3430.14673	263.85744		
Corrected Total	14	14661			

Root MSE	16.24369	R-Square	0.7660
Dependent Mean	29.49800	Adj R-Sq	0.7480
Coeff Var	55.06709		

Parameter Estimates						
Variable	Label	DF	Parameter Estimate	Standard Error	t Value	Pr > |t|
Intercept	Intercept	1	1.98831	5.94734	0.33	0.7435
gold_prices	gold_prices	1	0.05965	0.00914	6.52	<.0001

Figure 7-4. *Output*

Figure 7-4 shows that the p-value is < 0.0001, which means that we have enough evidence to reject the null hypothesis, and that there is a linear relationship between oil and gold prices. The R-Square value represents how strong that relationship is. The slope of the line = 0.05965. From the Parameter Estimate, the equation of the line is:

```
oil_price = 1.98831 + 0.05965 gold_prices
```

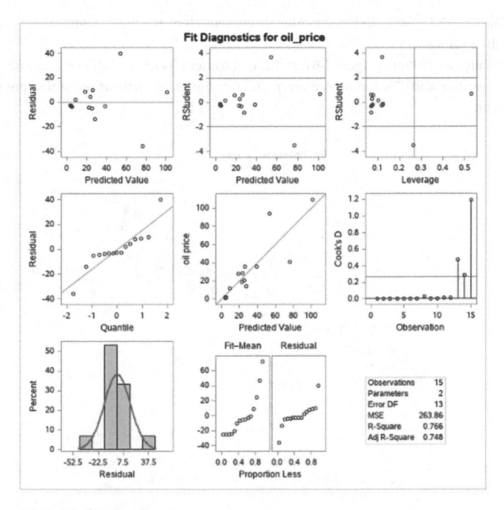

Figure 7-5. *Fit diagnostics*

Figure 7-5 shows the fit diagnostics for the oil_price variable. Residual shows a random pattern of dots above and below the 0 line, which indicates an adequate model.

The RStudent by Predicted Value plot shows a couple of dots outside the ±2 limits. A third dot lies in between the two lines but is far away from the rest of the dots.

The RStudent by Leverage plot shows the outliers that have leverage on the calculation of the regression coefficients.

The Residual by Quartile plot again shows those three points.

The Dependent Variable (TASK) by the Predicted Value plot visualizes variability in the prediction. In this plot, the dots are random, and there is not a pattern that indicates a constant variance of the error.

The Cook's D plot is designed to identify outliers. Here, the three points are very clear with their values.

Finally, the FitPlot in Figure 7-6 consists of a scatter plot of the data overlaid with the regression line and 95% confidence and prediction limits. Again the three outlier points are clear.

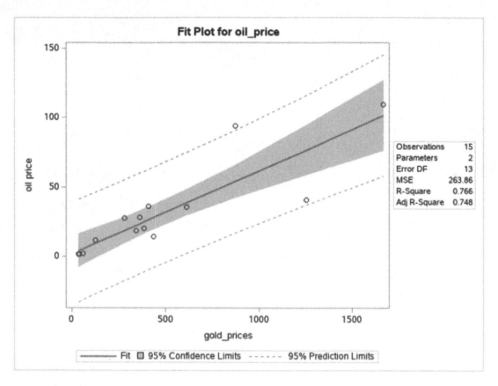

Figure 7-6. *Fit plot*

Multiple Linear Regression

In multiple linear regression, there is more than one independent variable in the model. The line equation will be as follows:

$$Y = a + b_1 X_1 + b_2 X_2 + \ldots + b_i X_i$$

Where Y is the dependent variable for which we want to predict its value, a is the intercept, b_i are the coefficients, and X_i are the independent variables. The error is ignored.

The null hypothesis is that there is no linear relationship between the dependent variable and any of the independent variables. In other words, the values of b_i are zeros. The alternative hypothesis is that there is a linear relationship, and any b_i does not equal zero.

H0: $b_i = 0$

Ha: $b_i \char`\^= 0$

Let us see if there is a linear relationship between the stock performance index (S & P index) and the prices of oil per barrel and gold in presidential election years.

We shall use the same dataset as in the previous example of the simple linear regression. The dataset is in the "Dataset" folder and has the name sp_oil_gold.xlsx. From SAS Studio, upload it and click Refresh, and then import it, as we explained in Chapter 2. Change the name of the dataset from IMPORT to sp_oil_gold.

Let us first explore the three variables. Run the code found in the Listing 7-1.sas file in the "Example Code" folder. Listing 7-1 displays three vertical bar charts showing the prices of oil, gold, and stocks over the years (from 1960 to 2016), each in a dedicated graph. The years are the presidential elections years.

Listing 7-1. Exploring the Stock, Oil, and Gold Prices

```
ods graphics / reset width=6.4in height=4.8in imagemap;
proc sgplot data=  WORK.SP_OIL_GOLD noautolegend ;

/*this vertical bar chart will show the stock performance over the years*/
vbar Year / response=Stock_Market_Returns  group=Stock_Market_Returns
groupdisplay=cluster  datalabel;

yaxis grid;

run;

ods graphics / reset;
ods graphics / reset width=6.4in height=4.8in imagemap;

proc sgplot data=  WORK.SP_OIL_GOLD noautolegend;

/*this vertical bar chart will show the oil prices over the years*/

vbar Year / response=oil_price group=oil_price groupdisplay=cluster datalabel;

yaxis grid;
```

```
run;

ods graphics / reset;
ods graphics / reset width=6.4in height=4.8in imagemap;

proc sgplot data=  WORK.SP_OIL_GOLD noautolegend ;

/*this vertical bar chart will show the gold prices over the years*/
vbar Year / response=gold_prices group=gold_prices groupdisplay=cluster
 datalabel;

yaxis grid;

run;

ods graphics / reset;
```

The output will be as shown in Figures 7-7, 7-8, and 7-9. The 2008 financial crisis is clear in all the graphs. This crisis is the worst economic disaster in the United States since the Great Depression of 1929. Oil and gold prices spiked in 2012.

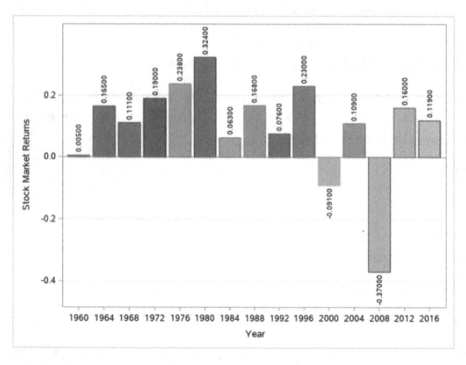

Figure 7-7. *Exploring S & P Index*

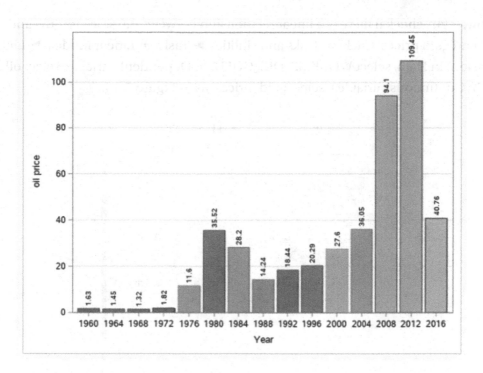

Figure 7-8. *Exploring oil price*

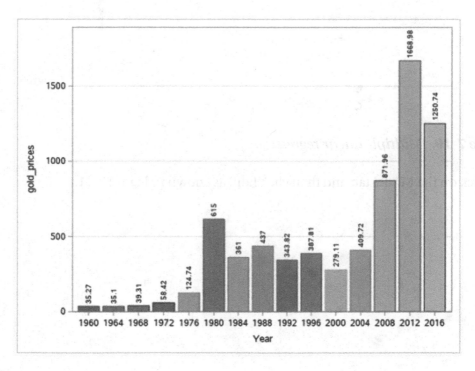

Figure 7-9. *Exploring gold price*

Now, let us check if there is a linear relationship between the stock returns and the oil and gold prices. Click on Tasks and Utilities ➤ Tasks ➤ Linear Models ➤ Linear Regression. In Data, select WORK.SP_OIL_GOLD. In Dependent variable, select oil_ price. In Continuous variables, select gold_prices, as in Figure 7-10.

Figure 7-10. *Multiple linear regression*

Click on the Model tab and then click Edit, as shown in Figure 7-11.

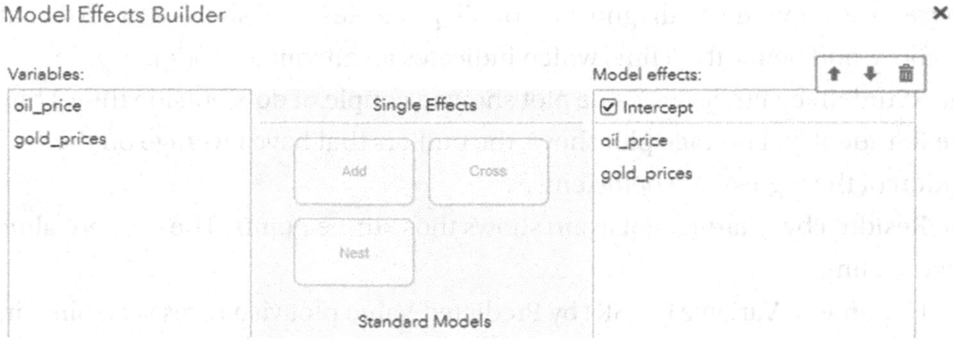

Model Effects Builder   ✕

Variables: Model effects: ↑ ↓ 🗑

oil_price   Single Effects   ☑ Intercept

gold_prices   oil_price

                              Add   Cross   gold_prices

                              Nest

                              Standard Models

Figure 7-11. *Model tab*

Click Run. The output is shown in Figure 7-12. Figure 7-12 shows that the p-value is < 0.05, which means that we have enough evidence to reject the null hypothesis, and that there is a linear relationship between the stock returns and the oil and gold prices. The R-Square value represents how strong that relationship is. From the Parameter Estimate, the equation of the line is as follows:

```
stock_market_returns  = 0.13010 - 0.00671 oil_price + 0.00036367 gold_prices
```

Model: MODEL1
Dependent Variable: Stock_Market_Returns Stock Market Returns

Number of Observations Read	15
Number of Observations Used	15

Analysis of Variance					
Source	DF	Sum of Squares	Mean Square	F Value	Pr > F
Model	2	0.15886	0.07943	4.40	0.0369
Error	12	0.21673	0.01806		
Corrected Total	14	0.37558			

Root MSE	0.13439	R-Square	0.4230
Dependent Mean	0.09980	Adj R-Sq	0.3268
Coeff Var	134.65846		

Parameter Estimates						
Variable	Label	DF	Parameter Estimate	Standard Error	t Value	Pr > |t|
Intercept	Intercept	1	0.13010	0.04942	2.63	0.0219
oil_price	oil price	1	-0.00671	0.00229	-2.93	0.0127
gold_prices	gold_prices	1	0.00036367	0.00015638	2.33	0.0384

Figure 7-12. *Output*

Figure 7-13 shows the fit diagnostics for oil_price. Residual shows a random pattern of dots above and below the 0 line, which indicates an adequate model.

The RStudent by Predicted Value plot shows a couple of dots outside the ±2 limits.

The RStudent by Leverage plot shows the outliers that have leverage on the calculation of the regression coefficients.

The Residual by Quartile plot again shows those three points. The dots are almost on the diagonal line.

The Dependent Variable (TASK) by Predicted Value plot visualizes variability in the prediction. In this plot, the dots are random, and there is not a pattern that indicates a constant variance of the error.

The Cook's D plot is designed to identify outliers. Here, the three points are quite clear with their values.

The Percent-Residual shows a normal distribution curve.

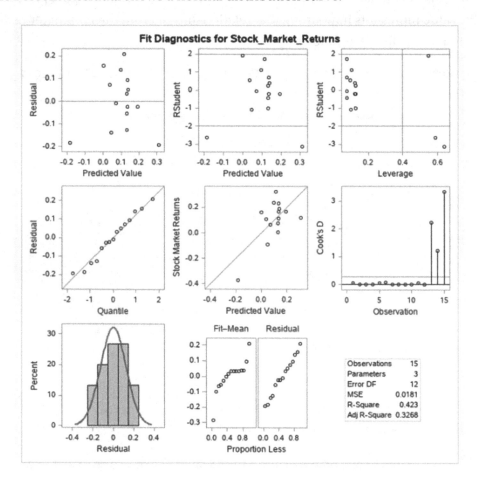

Figure 7-13. *Fit diagnostics*

Logistic Regression

Logistic regression is similar to linear regression except that the dependent variable is of binary values, not continuous. The model equation checks the probability of one of the discrete binary values of the dependent variable occurring.

The binary logistic regression model equation is as follows:

$$p = e\wedge (b_0 + b_1 X_1 + b_2 X_2 + ...) / (1 + e\wedge (b_0 + b_1 X_1 + b_2 X_2 + ...))$$

Again, the null hypothesis is that there is no relationship between the variables, so all the coefficients are zeros. The alternative hypothesis is that there is a relationship between the dependent variable and the independent variables.

For this example, we use the Titanic dataset to see if there is a relationship between surviving and age and passenger's class as independent variables.

The dataset is in the "Dataset" folder and has the name logistic_titanic_passengers. csv, or you can download the CSV file from OpenDataSoft.com at the following URL: https://public.opendatasoft.com/explore/dataset/titanic-passengers/export/.

From SAS Studio, upload the file and click Refresh, and then import it, as we explained in Chapter 2. Change the name of the dataset from IMPORT to TITANIC.

Click on Tasks and Utilities ➤ Tasks ➤ Linear Models ➤ Binary Logistic Regression. In Data, select WORK.TITANIC. In Response variable, select Survived. For the Event of Interest, select Yes to see the probability of the survived people, not the deceased. In Explanatory variables, select Age and Pclass for Continuous variables, as in Figure 7-14.

Figure 7-14. *Binary logistic regression*

Click on the Model tab, select Custom Model, and click Edit, as in Figure 7-15.

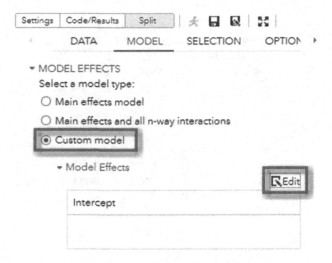

Figure 7-15. *Model tab*

Select Age and Pclass and click Add, and then scroll down and click OK, as in Figure 7-16.

Model Effects Builder

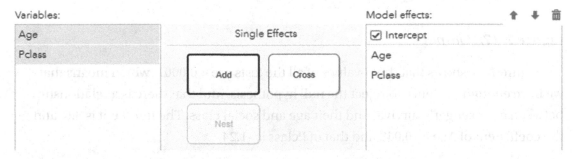

Figure 7-16. *Model effects builder*

Click Run. The output is shown in Figure 7-17.

Testing Global Null Hypothesis: BETA=0			
Test	Chi-Square	DF	Pr > ChiSq
Likelihood Ratio	137.0865	2	<.0001
Score	128.8185	2	<.0001
Wald	110.0444	2	<.0001

Analysis of Maximum Likelihood Estimates					
Parameter	DF	Estimate	Standard Error	Wald Chi-Square	Pr > ChiSq
Intercept	1	3.5853	0.4068	77.6945	<.0001
Age	1	-0.0420	0.00672	39.0142	<.0001
Pclass	1	-1.2438	0.1191	109.1414	<.0001

Odds Ratio Estimates		
Effect	Point Estimate	95% Wald Confidence Limits
Age	0.959	0.946 0.972
Pclass	0.288	0.228 0.364

Association of Predicted Probabilities and Observed Responses			
Percent Concordant	74.1	Somers' D	0.489
Percent Discordant	25.2	Gamma	0.492
Percent Tied	0.7	Tau-a	0.236
Pairs	122960	c	0.744

Figure 7-17. *Output*

Figure 7-17 shows that the p-values of all the tests are < 0.0001, which means that we have enough evidence to reject the null hypothesis, and that there is a relationship between a passenger's survival and their age and social class. The intercept is 3.6, and the coefficient of Age is -0.042 and that of Pclass is -1.24.

Summary

This chapter explains the single, multiple, and logistic regressions. It explains in detail the diagnostic matrix plots and the meaning of each of them. Moreover, it shows how to easily spot the outliers in several plots of the diagnostic plots. In the next chapter, we shall talk about the most popular SAS product on the market nowadays, which is SAS Viya. SAS Viya is an all-in-one product. Therefore, it includes SAS Studio in it as well. We shall touch on some features and options that are not available in the free version of SAS Studio and are added value to the SAS Studio packaged with SAS Viya. Moreover, we shall do a report using SAS Visual Analytics.

CHAPTER 8

SAS Visual Statistics: Viya

The free version of SAS Studio does not contain all of the available features. Therefore, in this chapter, we will see some advanced features that are available in SAS Viya. SAS Viya is one of the most popular products of SAS, and its platform includes SAS Studio in it.

About SAS Studio, SAS Visual Statistics, SAS Visual Analytics, and SAS Viya

Too many names, and you might get confused about which does what, right?

SAS Visual Statistics is an add-on to SAS Visual Analytics, which is a product that has been built on top of the Viya platform. Hence, they are all a stack built upon each other.

SAS Visual Analytics enables you to explore data, apply predictive analytics, and build interactive reports. SAS Visual Statistics extends these capabilities by creating, testing, and comparing models based on the patterns discovered in SAS Visual Analytics.

SAS Viya is a cloud-based product. I think that it was built with the philosophy of having one solution for all users. From data scientists to business analysts, application developers to executives, anyone can use its point-and-click interface. The programming interface of SAS Viya is SAS Studio. However, SAS Viya can run code from other programming languages, such as Python, R, and Java.

Several products have been built on top of the Viya platform: SAS Visual Analytics, SAS Visual Statistics, SAS Visual Investigator, and SAS Visual Data Mining and Machine Learning.

If you subscribed to try any of these products, you would have access to the rest of them, as they are all integrated into the SAS Viya platform. However, in this chapter, we will focus merely on SAS Visual Statistics.

If you would like to know more about SAS Viya, please check out its big-picture video at this URL:

```
https://video.sas.com/detail/videos/sas-viya_/video/4911667600001/
sas%C2%AE-viya%E2%84%A2:-the-big-picture
```

© Engy Fouda 2020
E. Fouda, *Learn Data Science Using SAS Studio*, https://doi.org/10.1007/978-1-4842-6237-5_8

If you would like to watch a brief introduction video about SAS Visual Statistics, please check out the following URL:

https://players.brightcove.net/1872491364001/default_default/index.html?vid
eoId=5772778452001

To try SAS Visual Analytics and SAS Visual Statistics for free for fourteen days (with the option to extend if needed), please fill in this form:

https://www.sas.com/en_us/trials/software/visual-analytics-visual-
statistics/form.html

You will receive an email with instructions and links to your portal. No installation is needed at all, because SAS Viya is a cloud-based platform. On SAS Viya, there is an option to extend your trial period.

SAS Viya Tour

After you sign in to your Visual Statistics account on Viya, you will have an interface, as shown in Figure 8-1. The interface is divided into three panes. The contents pane on the left enables you to work with your data and select your model object type. The canvas in the middle is where you build your models. The properties pane on the right enables you to set model roles, options, and filters.

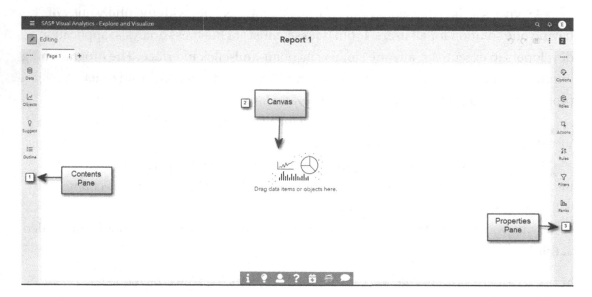

Figure 8-1. *Visual Statistics interface*

Later in this chapter, we shall discuss SAS Visual Statistics in more detail. Now, let us see the other products that are available inside SAS Viya along with SAS Visual Statistics. Click on the three lines at the top left to show the applications menu, as in Figure 8-2.

Figure 8-2. *Click on the three lines to show the applications menu*

After clicking, all the available applications will show, as in Figure 8-3. In the next section, we shall use SAS Studio on the Viya platform by clicking on "Develop SAS Code."

Figure 8-3. *SAS Viya applications*

Map of Counties

We will use SAS Studio on SAS Viya to create a map of selected counties within a state with PROC GMAP.

We modify the code of the example listed at the following URL to visualize Democratic and Republican counties in the state of Maine:

http://support.sas.com/kb/24/898.html

From the following URL, Maine's FIPS Code in SAS is 23:

https://documentation.sas.com/?cdcId=sasstudiocdc&cdcVersion=5.1&docsetId=g rmapref&docsetVersion=9.4_01&docsetTarget=n1w5dpp87y5qxon1vxqh3bs4m444.htm& locale=en&activeCdc=pgmsascdc

FIPS stands for Federal Information Processing Standards. Here are the FIPS codes for Maine counties: https://en.wikipedia.org/wiki/List_of_counties_in_Maine

Listing 8-1. Political Parties Illustrated on Maine's Counties for 2016 Presidential Election

```
* Political Parties Illustrated on Maine's Counties for 2016 Presidential
Elections;
/* Set the graphics environment */
goptions reset=all cback=white border htitle=12pt htext=10pt;

 /* Create a sample data set */
data voters;
/* input the FIPS values of the state and the counties, then the number of
voters who voted for the democratic candidate and the number of voters who
voted republican */
    input State County Republican Democrat;

/*if the county's number of democratic voters are higher than the
republicans then mark this county's party as 1 and 2 if republicans are
higher*/
if democrat > republican then party=1;
else if democrat< republican then party=2;
```

```
/*actual data values*/
   datalines;
23 1 28189 22975
23 3 19419 13377
23 5 57697 102935
23 7 7900 7001
23 9 13682 16107
23 11 29296 31753
23 13 9148 12440
23 15 9727 10241
23 17 12172 16214
23 19 41601 32832
23 21 5403 3098
23 23 9304 10679
23 25 14998 9092
23 27 10378 10442
23 29 9037 6358
23 31 50388 55828
;
run;

/* Project the map coordinates and create a new map data set */
proc gproject data=maps.counties out=newmap;
where state=23 ;
id state county;
run;

/*the value pattern is solid and the color is blue or red*/
pattern1 v=s c=blue;
pattern2 v=M3X0 c=red;

/* Generate the map */
proc gmap data=voters map=newmap all;
   id state county;
```

```
    choro party / nolegend ;
    title "Political Parties Illustrated on Maine's Counties for 2016
    Presidential Election";

run;
quit;
```

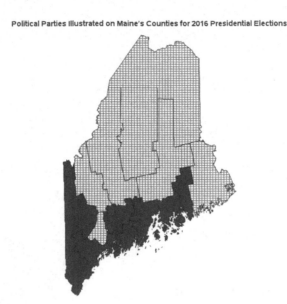

Political Parties Illustrated on Maine's Counties for 2016 Presidential Elections

Figure 8-4. *Political parties illustrated on Maine's counties for 2016 presidential election*

Listing 8-1 starts by creating a new dataset called voters that contains four columns: state, county, democrat, and republican. The first column, state, is the state's FIPS code, which is 23, indicating Maine. The second column, county, is the county's FIPS code. The third column, democrat, is the number of voters who voted for the Democratic candidate, while the last column, republican, is the number of voters who voted for the Republican candidate.

Then, the code compares the Democratic and Republican numbers of voters. If the Democratic is larger, this county's party is 1, else it's 2. The code uses these numbers to color the county. The datalines section has the actual values to fill in the dataset.

The new section uses PROC GPROJECT to project a new map that has only the state of Maine and its counties by using the FIPS values.

Then it sets two patterns (Figure 8-4). The first one is for party=1, which is solid blue for Democrats, and the second is M3X0, which is crosshatched lines, with red color for party=2, for Republicans. You can check the SAS documentation for further details about the PATTERN keyword at this URL: `https://documentation.sas.com/?docsetId=graphref` `&docsetTarget=p02daopdgx5yh4n1vmffk4ud4697.htm&docsetVersion=9.4&locale=en`.

PROC GMAP uses these two patterns to generate the colored choro map using the voters dataset and the map that PROC GPROJECT generated. To learn more about PROC GMAP, please check the following URL of the SAS documentation:

`https://documentation.sas.com/?docsetId=grmapref&docsetTarget=n0ncszzwvnx84` `bn1ql4487r08jpa.htm&docsetVersion=9.4_01&locale=en`.

To verify the results, I checked plenty of sites to compare my results with theirs. A couple of them are as follows:

- `https://www.politico.com/2016-election/results/map/` `president/maine/`

- `https://en.wikipedia.org/wiki/2016_United_States_` `presidential_election_in_Maine`

Similarly, I was able to compare the parties' distribution over the counties over the years, which was mentioned in Chapter 1, in Figure 1-12.

I added the code for the years 2012, 2008, 2004, and 2000 to the "Example Code" folder. They are Listing 8-2.sas, Listing 8-3.sas, Listing 8-4.sas, and Listing 8-5.sas. These listed programs generate the maps in Figure 8-5.

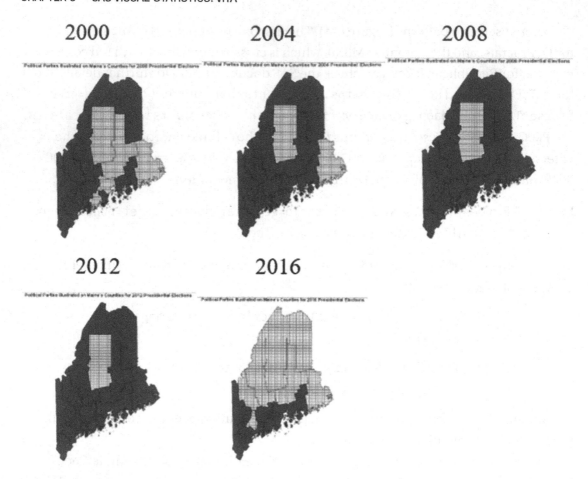

Figure 8-5. *Maps of how Maine's counties voted in the presidential elections in years 2000–2016*

For further resources, please check the appendix.

SAS Visual Statistics: First Report

Now, in order to return to Visual Statics to compile our first interactive report, we need to upload our dataset. In this section, we use the Accidental Drug-related Deaths in the State of Connecticut dataset.

You can download the CSV from the following link:

```
https://catalog.data.gov/dataset/accidental-drug-related-deaths-january-
2012-sept-2015
```

However, we only use the data from 2016 to 2018. You will find the dataset in the "Datasets" folder with the name Accidents_Drugs_CT.csv. Click on the three lines, as in the previous section; select "Manage Data," as in Figure 8-6.

Figure 8-6. *Manage data*

Click Import ➤ Local File, as in Figure 8-7.

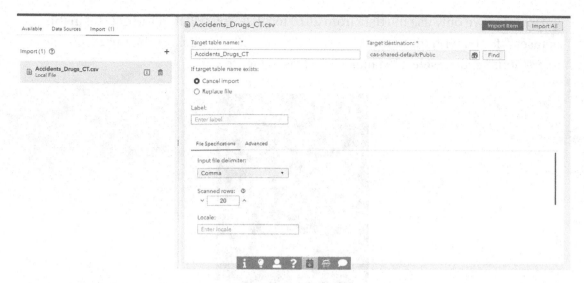

Figure 8-7. *Import the dataset of accidental deaths associated with drug overdose in Connecticut in years 2016–2018*

Then, click on Import Item. Then, click on the three lines, and then click Explore and Visualize, as in Figure 8-8.

Figure 8-8. *Explore and Visualize*

On the right pane, click on Options, as in Figure 8-9, to change the report's background color and font. Let us go for a darker gray background and yellow font, as in Figure 8-9.

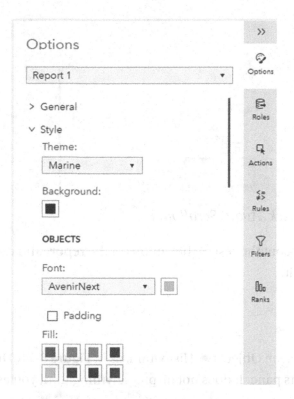

Figure 8-9. *Enhance the report background*

Scroll down the options to the Layout section and uncheck "Avoid Scrollbars," as in Figure 8-10. This is crucial, as when you leave the default option, the plots and reports are not displayed clearly. Moreover, it locks the resizing features of the report sections. Hence, it is better to uncheck this option to be able to control the dimensions of your report.

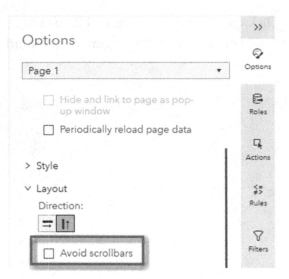

Figure 8-10. *Uncheck "Avoid Scrollbars"*

Feel free to check out the rest of the options of the report and explore how to enhance the look of it.

Histogram

On the left pane, click on Objects ➤ Histogram, as in Figure 8-11. Drag it onto your report in the contents pane. It does not display anything until you assign a role for it, as in the next step.

Figure 8-11. *Histogram*

On the right pane, click on Roles to assign the Measure, as in Figure 8-12. For histograms, only numeric values are allowed. Hence, Age is the only feasible variable in this dataset.

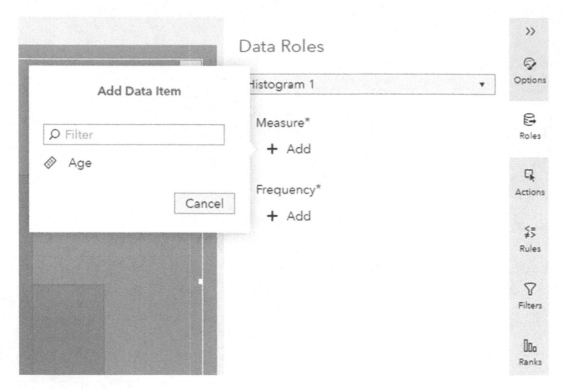

Figure 8-12.   *Only numeric values are allowed as a histogram's measure*

SAS Visual Statistics generates a plot, as in Figure 8-13, showing that the highest age range that experienced accidental drug-related deaths in the state of Connecticut between the years 2016 and 2018 is 31–34 years old.

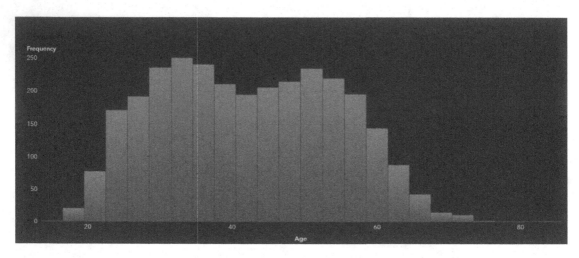

Figure 8-13.   *The age distribution of people who died with drug overdose in Connecticut in years 2016–2018*

Note   To see Figures 8-13, 8-14, and 8-15 in color, please access this book's source code.

To enhance the appearance of the title, double click on it, and a menu pops up, as in Figure 8-14. Change the color, and make it bold and centered.

Figure 8-14. *Enhance the title appearance*

As you can see in Figure 8-15, the title is more clear now.

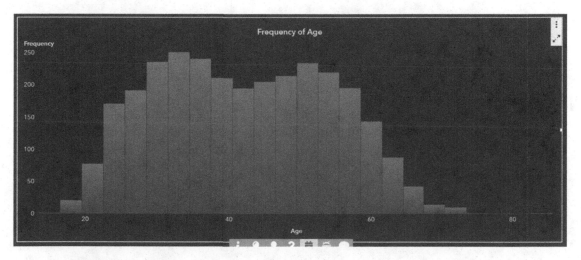

Figure 8-15. *The final look of this part of the report*

Word Cloud

The word cloud is an exciting feature of the natural language processing of character variables. It displays the most frequently used words in any variable larger than the rest of the words. Again, from the right pane, click on Objects ➤ Word Cloud, as in Figure 8-16. Drag it to the report.

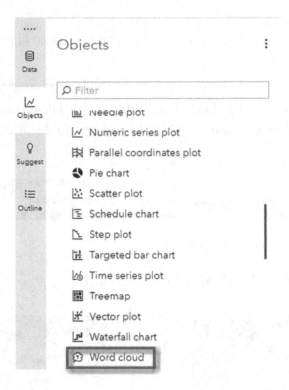

Figure 8-16. *Word cloud*

Assign a role to it from the right pane. Select Roles, and in Word, select DeathCity, as in Figure 8-17.

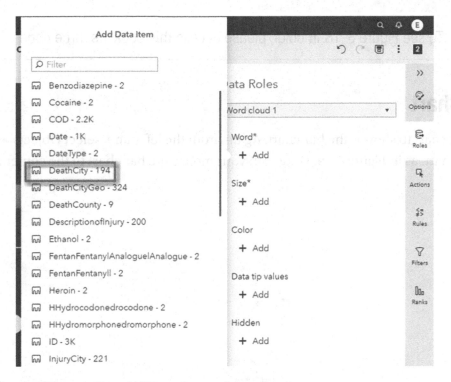

Figure 8-17. *Choose DeathCity column from the dataset*

The word cloud, as in Figure 8-18, shows that Hartford is where these lethal drug-related deaths most often happen, followed by New Haven, Bridgeport, and Waterbury. This cloud can help law enforcement to increase surveillance in these cities. Furthermore, you can add a plot of the most recurring dates and times to check if there is a relationship between certain holidays or times and these incidents. You can dig more by tracking down the streets to identify if there are certain locations that sell drugs in these cities.

Figure 8-18. *Cities that have higher numbers of drug-related mortalities are in a larger font than others*

Note   To see Figure 8-18 in color, please access this book's source code.

Bar Chart

Another chart to show is the bar chart. Again, from the left pane, select Objects ➤ Graphs ➤ Bar Chart, as in Figure 8-19. Drag with your mouse the bar chart to the contents pane.

Figure 8-19. *Bar chart*

After dragging it, it will write you a note that there is no role assigned. From the right pane, select Roles. For Category, select Race, as in Figure 8-20.

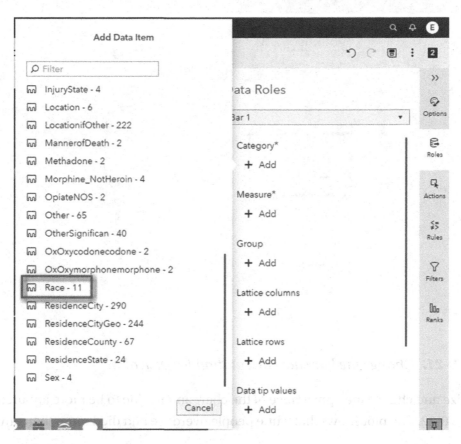

Figure 8-20. *Select Race from the dataset*

By default, the bar chart in SAS Visual Statistics is horizontal. On the right pane, click on Options ➤ Bar. Click on Vertical direction, as in Figure 8-21.

Figure 8-21. *Change the bar chart orientation to vertical*

Resize and change the appearance of the chart and its title to be more apparent, as in Figure 8-22. The plot shows that white people overdose and died more than any other race.

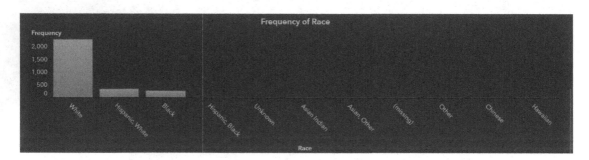

Figure 8-22. *The race distribution of overdose deaths*

Note   To see Figure 8-22 in color, please access this book's source code.

Butterfly Chart

A butterfly chart displays two bar charts with a shared category axis to compare between them. The baselines of the two bar charts are located in the center of the chart. Click on Objects ➤ Graphs ➤ Butterfly Chart, as in Figure 8-23. Drag it onto your report under the previous charts or beside any of them.

Figure 8-23. *Butterfly chart*

On the right pane, select Roles, and in Category, select Sex, as in Figure 8-24. In Measure (bar), choose Age, and in (bar2), select Frequency, as in Figure 8-24.

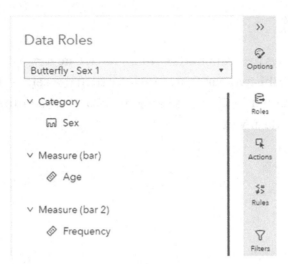

Figure 8-24. *Butterfly chart will show which gender overdosed more*

Figure 8-25 shows that males commit such deadly drug overdoses more than females.

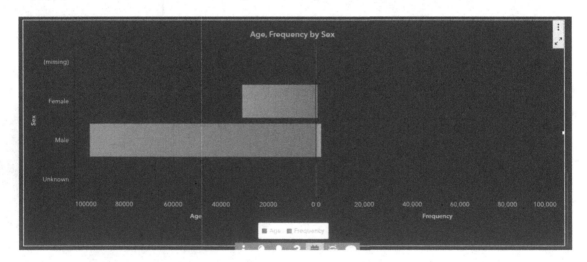

Figure 8-25. *Males overdose a great deal more than do females*

Summary

This chapter introduced SAS Viya, which is an all-in-one product. It includes SAS Studio in it as well. We talked about the choropleth maps, which are not available in the free version of SAS Studio. Moreover, we created a report that contained many infographics in one sheet. It has some interesting features, such as the word cloud. In the next chapter, we shall talk about where to go after this book and how to continue your track in learning data science using SAS and how to hunt yourself down some data science jobs.

CHAPTER 9

What Is Next?

As we discussed in Chapter 1, the data science process starts with a question or hypothesis. Then, you collect as much relevant raw data as you can, then clean and explore it. The fourth stage is modeling and evaluating. The last step is deploying, visualizing, and communicating results in reports, as in Figure 9-1.

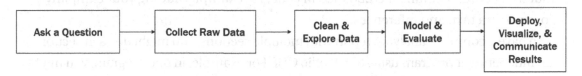

Figure 9-1. *Data science process*

It is worth mentioning that the first three phases in the data science process are called data analytics, while the modeling and evaluation phase is data science. The visualization and reporting of the results are common between data science and analytics.

Therefore, the data analytics cares more about analyzing the real historical data that we already have, while data science's scope is predicting the future. In this book, we presented the modeling and evaluation phase in Chapters 7 and 8. Moreover, we discussed the difference between supervised and unsupervised algorithms in Chapter 3.

However, there are plenty of other data science algorithms that you should know about as a data scientist; for example, decision trees, support vector machines, and neural networks. If you are interested in image analysis, then go ahead to learning about deep learning and TensorFlow.

© Engy Fouda 2020
E. Fouda, *Learn Data Science Using SAS Studio*, https://doi.org/10.1007/978-1-4842-6237-5_9

Reports: How Do You Present Your Results?

Throughout the book, we have talked in several chapters about reports. In this section, we assemble these scattered puzzle pieces to learn how easy it is to write a report in SAS Studio in comparison to other languages.

In SAS Studio, compiling a report is much easier than in other data science languages. For example, in R, to compile a report that collects all your graphs, statistical analyses, and inferences, you need to use Markdown to generate an HTML document at the end. Adding your companies' logos and following its report templates might be troublesome to some newbies to data science if they are not familiar with HTML code. While in SAS Studio, as we mentioned in Chapter 2, Figure 2-7, the report output can be in RTF, PDF, or HTML formats. Editing the RTF file in any word processing application is much easier than editing it embedded in code; for example, adding your company's logo, any extra text, and inferences.

To have a complete analytics report with multiple sections, you might do each section as an independent program using SAS Studio GUI. For example, in one program, you might create the statistical analysis using one-way frequencies. In another program, you might do the histogram. In a third program, you might make a visualization using the maps, and so on. In the end, you would want to combine all these programs in one to generate one report only. You can do that by clicking on the Edit tab, as in Figure 3-39. At each of the programs that we mentioned, copy the auto-generated code and paste after each other in one program. Run this new, large program and compile your report from it. We did that in Listing 7-1, where we explored the oil, the gold, and the stocks. We copied the three programs and compiled them in one report that had all three charts; then, we performed the regression.

In conclusion, by copying the scattered code from several programs, you can generate one report. Then, click on Print and save your report in RTF format and edit it in any word processing application. In SAS Studio, there is no need to write special code to generate the report.

How Do You Write Your Results, Keeping Your Audience in Mind?

After we are done with the technicalities of generating the report from SAS Studio, we need to think about our audience, who will read it. Our audience varies and includes data scientist peers, our team leaders, the company board of directors, the marketing team, and our company's clients.

For example, our clients hired us to decide whether to buy or sell a stock. They would not care about histograms or maps. They would want plain English to know the justification for our decision. On the other hand, our manager and team will need to know every detail and need even to be able to read and understand the code.

It is crucial to know the deliverables required for every entity and write our results so as to be customized to our audience's needs. Explaining the visualizations is more critical than the visualizations themselves.

Here are some further links for data analysis reports examples:

- `https://www.datapine.com/blog/analytical-report-example-and-template/`

- `http://www.stat.cmu.edu/~brian/701/notes/paper-structure.pdf`

- `https://www.examples.com/business/data-analysis-report-examples.html`

- `https://bizfluent.com/how-5822409-write-good-data-analysis-report.html`

Here are links for data science report structure:

- `https://www.kaggle.com/WinningModelDocumentationGuidelines`

- `https://github.com/rdpeng/courses/blob/master/05_ReproducibleResearch/Checklist/Reproducible%20Research%20Checklist.pdf`

How Can You Make Money Online from Data Science?

You can find a vast number of articles about how to make money online from data science. In this section, I share with you a slice of my experience and what I actually do. I am a freelance instructor, consultant, journalist, and author.

Teaching Online

I have been teaching computer and technical courses my whole life. I taught my first course when I was still an undergraduate in my second year at the Faculty of Engineering, Cairo University. In Egypt, studying engineering takes five years, not four. I became a Microsoft Certified Professional in my fourth year. After graduating, I passed many of Microsoft's tracks and became a Charter at .NET track and a Microsoft Certified Trainer.

Yes, getting certified advances your career and proves your expertise. Others prefer showing their skills by making online tutorials, blogging, and answering others' questions on technical communities. I always read and hear about people who make thousands of dollars from YouTube. However, I do not know anyone who actually does that for a living.

Anyway, that other way around would not have worked for me, because to teach technically at certified centers, you must be certified. However, the con to that is that you must keep chasing the newest technologies, getting certified to always be top-notch. Otherwise, you lose your sparkle. If you get tired at any instant, you can switch to teaching basic sciences and technologies academically at universities. At some phases of my life, I did that as well, but soon I missed the adrenaline and switched back to technology instead of academia. Of course, teaching technically is much more rewarding financially than teaching at any university. Currently, I teach SAS, Docker, and Kubernetes tracks as a freelance instructor at ONLC Training Centers, Microsoft Partners, and at other certified centers.

I started exploring data science in a similar way. My first step in the field was to get certified by taking the Harvard Graduate Professional Certificate. It was an exhilarating experience. I shall talk to you later about it in detail because I cannot degrade it to a paragraph in a section.

Writing Online

I write online for several venues as a freelancer and as a volunteer. Volunteering usually brings me job or project offers. It is another way to show your experience. However, you must do volunteering with sincerity. I never thought of my volunteering as a chance to show my experience, but in many cases, I received job offers because of it.

Once, I wrote an article for free at *Medium* merely to share my knowledge and experience after I had a disagreement with my editor at one of the outlets. I was surprised to receive an offer from *Medium* to join their partner program. After publishing with the program, I started receiving money and another offer to write at one of the *Medium* magazines. The advantage of this program is that you are not compelled to write about certain topics or to have an editor who might reject your article. You can write whenever you have time about whatever you like and earn money if people read and like your articles. Here is the link to join the *Medium* partner program:

```
https://medium.com/creators
```

Although I write for several venues for larger fees than *Medium*, I prefer it the most. I vouch for it as the best venue for which to write whenever you have a chance. Writing regularly is the best practice to get a steady flow of cash and followers. Unfortunately, I do not do that yet, because I am mostly busy teaching. I consider writing as another way of teaching. However, I prefer the human interaction of teaching. Others of my friends prefer writing over teaching because articles last longer.

Kaggle-Zillow Competition

Kaggle is a third way to make money online. I know some people resigned from their companies and earn their living by competing at Kaggle. It is a data science hub, where universities and companies post projects with datasets as challenges with cash prizes. I tried it once, and it was a wonderful experience; I learned a lot from it.

The competition I joined was for $1,200,000 over two rounds. The participants would compete for about a year and a half. After the first round, only 100 participants whose scores were most accurate would join the second round. There were small prizes after the first round of $50,000. After the second round, the first prize is one million dollars. The goal of the contest is to predict house prices for the coming two years. The competition was presented by Zillow.

The competitions' datasets are usually of two types: one for training and the other for testing. Moreover, some competitions are open in the number of kernel submissions, while others are restricted to the number of submissions. So, read the rules carefully so as not to waste your chances.

You can check the contests' file, prizes, and timelines at this link: `https://www.kaggle.com/c/zillow-prize-1`.

The timeline for the contest I entered started in May 2017 and ended in January 2019. Therefore, it is not easy to win contests. Personally, I do not recommend that you resign from your work to earn a living from competing at Kaggle unless you actually tried and learned the winning formula.

At Kaggle, the participants can collaborate and post their code or results if they like. This sharing raises the rank of the kernel and the discussion rank of the participant. Some kagglers get famous, and their contributions get plenty of thumbs-ups. Employers and research centers many times hire these popular Kagglers even if they do not win competitions. Hence, Kaggle is equivalent to GitHub in showing the data scientists' skills and programming styles.

Again, be careful when you submit your kernel as to whether it is public or private. Some competitions prohibit you from changing your kernel's privacy after you set it the first time. Some competitions accept sharing your kernel once and making another private.

For this competition, I used the XGBoost algorithm for predicting the prices. I learned about the XGBoost algorithm from Kaggle. So, it is an educational spot for beginners as well. Recently, Kaggle started offering free online data science courses and free datasets for trying. Here is the link: `https://www.kaggle.com/learn/overview`.

If you have an issue or a question, feel free to post on the discussion board. The Kaggle support team and the community are highly active, supportive, and cooperative. After some time, you can get to know other kagglers, form a team with them, and compete together.

My rank in the Zillow competition was 231 out of 3131 teams. My algorithm was working fine, and by altering some values, the results were enhancing tremendously. I jumped from position 1000 to 231 in the leaderboard in a couple of weeks. Unfortunately, I was unable to continue competing as I had to travel to Boston to attend a couple of courses on campus. You might disagree with my decision; each person prioritizes differently. To me, my master's degree at Harvard University was my highest priority and worth more than a million dollars.

Harvard University, Extension School: Data Science Graduate Professional Certificate

This certificate consists of four courses that you can choose to attend online or on campus. All the lectures and labs are live-streamed and recorded. You are free to watch them as many times as you need even after graduation. If you think about it and consider taking it to advance your career, I highly recommend that you start right away. The course prices increase every semester, so the faster, the cheaper. No application or deep prior experience is required. For many people I know, it was easy to enroll but too hard to finish. The certificate needs much work and practice. You cannot take it on the fly. You must earn at least B in all the courses and finish them before three years. The recommended time is eighteen months. One of the courses must be a statistics course, and another mandatory course is the data science course. The other two courses are electives.

I finished the certificate in about a year, and all of my grades were A and A-. For two of these courses, I earned not only the full mark but also the bonus marks. By the way, there is no A+ at Harvard University! I started with the statistics course as a summer condensed course at Harvard University, Summer School. The course instructor was Professor Michael Parzen. He is one of the best professors ever and is rated as the most popular professor at the university. In fall 2016, I took two courses in one semester. The one that I document through this book was with Professor Larry Adams. The other course was "Big Data in Health Care" with Professor Oleg Pianykh. I learned a lot from that course as well. In spring, I took the mandatory data science course with Professors Andrey Sivachenko and Victor Farutin. It was an outstanding course, indeed a magnificent experience.

I recommend that you start with the statistics course, then the mandatory data science, then take the electives. You do not always have the power to choose, because not all courses are offered every semester. It is always better to check with your school advisor. The advisors at Harvard University are awesome—so friendly and knowledgeable. Be aware that sometimes the rules or courses change after you enroll in any of the university's programs. Hence, it is always better to update your advisor with your progress and discuss with them the courses that interest you before registering. The advisor always checks if you will fulfill the certificate or degree requirements and will direct you to the best choices. My advisor, Allyson K. Boggess, and my enrollment coach, Kristin Keller, helped me tremendously. Definitely, I would not have finished my degree without their help and Professor Sallie Sharp's supervision as my capstone project director.

The certificate is expensive, but you will earn this money back in about a year. Having the Harvard University logo on your resume and LinkedIn account advances your job opportunities to a higher level. Besides the work and money, the Harvard experience changes your life and your way of thinking forever. Yes, it is tough; you will stay awake working and studying for days without an hour of sleep. However, trust me, you will love it!

Data science tracks, whether academic or technical, are ubiquitous nowadays. You can try them as well. Many universities other than Harvard University started offering degrees and certificates in data science.

SAS Certification

After finishing this book, you will have covered almost all the topics required for the following two certifications:

1. SAS Certified Specialist: Base Programming Using SAS

2. SAS Certified Advanced Programmer for SAS 9

The required topics for the Base Programming exam are:

- Read and create data files.

- Manipulate and transform data.

- Create basic detail and summary reports using Base SAS procedures.

- Identify and correct syntax and programming logic errors.

The required topics for Advanced Programmer are:

- Using advanced data step programming statements and efficiency techniques to solve complex problems.

- Writing and interpreting SAS SQL code.

- Creating and using the SAS MACRO facility. (This topic is the only one that is not discussed in this book.)

SAS Certification credentials are globally recognized as the premier means of validating your SAS knowledge and accessing higher-paying job opportunities. Of all skills considered by employers, SAS is one of the most valuable in terms of salary. Please check this link as well: `https://www.sas.com/en_us/careers.html`.

I am certified by Microsoft, Oracle, and Docker in tens of tracks. The safest and fastest way to get certified is to study from braindumps. You can buy them online. The braindumps are pools of actual exam questions and their correct answers. Do not memorize the answers because you might not get the same questions. Instead, research the questions that you do not know and try to understand their answers. Remember that in the end, at work, your knowledge will be tested and not your memory.

Stay in Touch

Please stay in touch. Let me know if I can help you in any way. I am always glad to give a hand.

My email: efoda@ieee.org.

My website: www.engyfoda.com.

My LinkedIn: https://www.linkedin.com/in/engyfouda/.

Summary

In this chapter, we talked about how to write your reports with your audience in mind. Then, we talked about the next steps in your data science career. You can do ahead to academic certificates or technical ones to prove your experience and work on real-world projects under mentoring. Moreover, we talked about how to make money online through your data science experience. Wish you best of luck!

Resources

SAS Documentation: Specifying Colors in SAS/GRAPH Programs
`https://documentation.sas.com/?docsetId=graphref&docsetT`
`arget=p0rclt5u06r7oen1puocgn9mrfh0.htm&docsetVersion=9.4`
`&locale=en`

SAS Help: CHORO Statement `http://support.sas.com/`
`documentation/cdl/en/graphref/65389/HTML/default/viewer.`
`htm#n0b7n2mnw4t6lhn0zoy86orv95z2.htm`

SAS Communities: Using Colors on Maps (gmap) `https://`
`communities.sas.com/t5/Graphics-Programming/using-`
`colors-on-maps-gmap/td-p/454367`

SAS Communities: How to Change Color on the Map and More
`https://communities.sas.com/t5/Graphics-Programming/`
`How-to-change-color-on-the-map-and-more/td-p/324124#`

SAS Help: PATTERN Statement: `http://support.sas.com/`
`documentation/cdl/en/graphref/63022/HTML/default/viewer.`
`htm#patternchap.htm#global-fig9`

Maine Zip Codes: `https://www.zip-codes.com/state/me.asp`

SAS Video Tutorials: Creating Visualizations: Using Network
Diagrams and Geo Maps `https://video.sas.com/detail/`
`video/4077902263001/creating-visualizations:-using-`
`network-diagrams-and-geo-maps`

© Engy Fouda 2020
E. Fouda, *Learn Data Science Using SAS Studio*, https://doi.org/10.1007/978-1-4842-6237-5

SAS Blogs: SAS Graphs for Presidential Elections `https://blogs.sas.com/content/sastraining/2012/11/05/sas-graphs-for-presidential-elections/`

SAS Blogs: Examining Voter Registration Data with SAS Visual Analytics `https://blogs.sas.com/content/sascom/2019/08/27/examining-voter-registration-data-with-sas-visual-analytics/`

SAS Blogs: How to Create Custom Regional Maps in SAS Visual Analytics 8.2 `https://blogs.sas.com/content/sgf/2018/01/26/how-to-create-custom-regional-maps-in-sas-visual-analytics-8-2/`

SAS Communities: US County-Level Map in SAS Visual Analytics `https://communities.sas.com/t5/SAS-Visual-Analytics/US-County-Level-Map-in-SAS-Visual-Analytics/td-p/250678`

Index

A

ANalysis Of VAriance (ANOVA), 121
Arithmetic operators
    calculation, 134
    missing value conventions, 134, 135

B

Bar chart
    ascending sort, 76
    creation, 72
    dataset, 74
    descending, 78
    HBar, 71
    histogram creation, 78–82
    reverse tick values, 78
    summary statistics, 73
    table selection, 75
    title, 72
Bubble chart
    editing code, 88
    final version, 88
    initial output chart, 84
    label, 85
    labels, 86
    SASHELP.CLASS library, 83
    sgplot function, 87
Bubble map
    creation, 92
    Data—NYC OpenData, 90

dataset, number of
    crimes, 91, 96
datasets folder, 89
enhance appearance, 95
Esri maps, 97, 98
number of crimes, 93, 94
NYC_crime_dim, 92

C

Cluster analysis
    agglomerative algorithm, 99
    dendrogram, 101
    divisive method, 100
    splitting, 100
    unsupervised learning
        algorithm, 98
    user interface, 98
Comment statement, 133
Comparison
    operators, 135, 136
Correlation
    analysis variable, 110
    data tab, 110
    GDP *vs.* voter turnout, 114
    horsepower *vs.* weight, 112
    options tab, 111, 114
    Pearson, 111, 115
    statistical procedure, 109
    voter turnout *vs.* GDP, 115

223

© Engy Fouda 2020
E. Fouda, *Learn Data Science Using SAS Studio*, https://doi.org/10.1007/978-1-4842-6237-5

Printed in the United States
By Bookmasters